BASIC OF ELECTRONICS

AUTHOR

Dr. Brijesh Iyer

Professor,

Assistant Professor (Sel. Gr.)

Department of E & TC Engineering

Dr. Babasaheb Ambedkar Technological University, Lonere

Maharashtra (India)

REVIEWER

Dr. Sanjay Nalbalwar

Professor & Head

Department of E & TC Engineering

Dr. Babasaheb Ambedkar Technological University, Lonere

Maharashtra (India)

KHANNA BOOK PUBLISHING CO. (P) Ltd.
4C/4344, Ansari Road, Darya Ganj, New Delhi, India

All India Council for Technical Education
Nelson Mandela Marg, Vasant Kunj, New Delhi, India

Dear Readers,

To prevent the piracy, this book is secured with HIGH SECURITY HOLOGRAM on the front title cover. In case you don't find the hologram on the front cover title, please write us to at contact@khannabooks.com or whatsapp us at +91-99109 09320 and avail special gift voucher for yourself.

Specimen of Hologram on front Cover title:

Moreover, there is a SPECIAL DISCOUNT COUPON for you with EVERY HOLOGRAM.

How to avail this SPECIAL DISCOUNT:

Step 1: Scratch the hologram

Step 2: Under the scratch area, your "coupon code" is available

Step 3: Logon to www.khannabooks.com

Step 4: Use your "coupon code" in the shopping cart and get your copy at a special discount

Step 5: Enjoy your reading!

ISBN: 978-93-5538-808-7

Book Code: UG408EN

Basic of Electronics

by Dr. Brijesh Iyer

First Edition: 2026

Originally Published by:

All India Council for Technical Education, Ministry of Education, GOI

Exclusively Published & Distributed by:

Khanna Book Publishing Co. (P) Ltd.

4C/4344, Ansari Road, Daryaganj,
New Delhi (INDIA)

Visit us at: **www.khannabooks.com**

Write us at: **contact@khannabooks.com**

To view complete list of books, Please scan the QR Code: KPH

Printed in Bharat.

प्रो. टी. जी. सीताराम
अध्यक्ष
Prof. T. G. Sitharam
Chairman

सत्यमेव जयते

आज़ादी का अमृत महोत्सव

अखिल भारतीय तकनीकी शिक्षा परिषद्
(भारत सरकार का एक सांविधिक निकाय)
(शिक्षा मंत्रालय, भारत सरकार)
नेल्सन मंडेला मार्ग, वसंत कुंज, नई दिल्ली–110070
दूरभाष : 011–26131498
ई–मेल : chairman@aicte-india.org

ALL INDIA COUNCIL FOR TECHNICAL EDUCATION
(A STATUTORY BODY OF THE GOVT. OF INDIA)
(Ministry of Education, Govt. of India)
Nelson Mandela Marg, Vasant Kunj, New Delhi-110070
Phone : 011-26131498
E-mail : chairman@aicte-india.org

FOREWORD

Engineers are the backbone of any modern society. They are the ones responsible for the marvels as well as the improved quality of life across the world. Engineers have driven humanity towards greater heights in a more evolved and unprecedented manner.

The All India Council for Technical Education (AICTE), have spared no efforts towards the strengthening of the technical education in the country. AICTE is always committed towards promoting quality Technical Education to make India a modern developed nation emphasizing on the overall welfare of mankind.

An array of initiatives has been taken by AICTE in last decade which have been accelerated now by the National Education Policy (NEP) 2020. The implementation of NEP under the visionary leadership of Hon'ble Prime Minister of India envisages the provision for education in regional languages to all, thereby ensuring that every graduate becomes competent enough and is in a position to contribute towards the national growth and development through innovation & entrepreneurship.

One of the spheres where AICTE had been relentlessly working since past couple of years is providing high quality original technical contents at Under Graduate & Diploma level prepared and translated by eminent educators in various Indian languages to its aspirants. For students pursuing 2nd year of their Engineering education, AICTE has identified 87 books, which shall be translated into 12 Indian languages - Hindi, Tamil, Gujarati, Odia, Bengali, Kannada, Urdu, Punjabi, Telugu, Marathi, Assamese & Malayalam. In addition to the English medium, books in different Indian Languages are going to support the students to understand the concepts in their respective mother tongue.

On behalf of AICTE, I express sincere gratitude to all distinguished authors, reviewers and translators from the renowned institutions of high repute for their admirable contribution in a record span of time.

AICTE is confident that these outcomes based original contents shall help aspirants to master the subject with comprehension and greater ease.

(Prof. T. G. Sitharam)

ACKNOWLEDGEMENT

The author is grateful to the authorities of AICTE, particularly Prof. T.G. Sitharam, Chairman; Dr. Abhay Jere, Vice-Chairman; Prof. Rajiv Kumar, Member-Secretary, Dr. Ramesh Unnikrishnan, Advisor-II and Dr. Sunil Luthra, Director, Training and Learning Bureau for their planning to publish the books on ***Basic of Electronics.***

I would like to express my sincere appreciation to Dr. Sanjay Nalbalwar, Professor & Head of the Department of E&TC Engineering at Dr. Babasaheb Ambedkar Technological University, Lonere-402103(MS)-India, for his contributions as a book reviewer. His efforts have made the book more accessible to students and have given it an artistic touch, resulting in a better overall shape.

I am deeply grateful to my wife, Dr. Prachi, and my children, Ku. Rajnandini and Chi. Prabhasdatt, for their unwavering patience throughout the completion of this project. Without their support, this work would not have come to fruition.

I would also like to acknowledge the reference books that have been instrumental in the preparation of this book: "Electronic Devices and Circuit Theory" by Robert Boylestad and Louis Nashelsky, "Digital Fundamentals" by Thomas Floyd, "Electronic Instrumentation and Measurements" by David Bell, and "A Course In Electrical and Electronic Measurements and Instrumentation" by A.K. Sawhney. Additionally, I extend my gratitude to Mr. Dixit Jain of Synergy Books Pvt. Ltd. Mumbai for his initial support and motivation.

Over the years, my students have made significant contributions that have enriched my experience and expertise in this subject. I am immensely grateful for their valuable input.

This book is the result of various suggestions provided by members of the All India Council for Technical Education (AICTE), experts, and authors who have shared their opinions and thoughts on advancing engineering education in our country. I extend my heartfelt acknowledgment to these contributors and all the individuals working in this field whose published books, review articles, papers, photographs, footnotes, references, and other valuable information have enriched our work during the writing process.

Dr. Brijesh Iyer

PREFACE

I take great pleasure in introducing the textbook "Basic of Electronics," which is the culmination of my extensive experience in teaching fundamental courses in electronics engineering. The motivation behind writing this book is to provide engineering students with a comprehensive understanding of the basic concepts and fundamentals of electronics engineering, allowing them to gain insight into the subject. With a focus on broad coverage and essential supplementary information, the book incorporates topics recommended by the All India Council for Technical Education (AICTE) in a systematic and organized manner. Special effort has been made to explain the fundamental concepts in the simplest possible way.

Throughout the preparation process, I extensively referred to various standard textbooks and developed sections such as critical questions, solved and supplementary problems, and more. Emphasis has been placed on definitions, laws, and comprehensive synopses of formulas for quick revision of the basic principles. The book addresses a wide range of medium and advanced-level problems, presented in a logical and systematic manner. These problem gradations have been tested over many years of teaching, catering to diverse student backgrounds.

In addition to relevant illustrations and examples, the book is enriched with solved problems in each unit to facilitate a comprehensive understanding of the topics. Each chapter is accompanied by solved examples, exercise questions, self-study questions, and multiple-choice questions.

I sincerely hope that this book will inspire students to delve into and discuss the underlying ideas behind the basic principles of electronics engineering. It aims to establish a solid foundation in the subject matter. I welcome and appreciate all valuable comments and suggestions that will contribute to the enhancement of future editions of this book. It is my utmost pleasure to present this book to both teachers and students. Working on the various aspects covered in this book has been a truly gratifying experience.

Dr. Brijesh Iyer

OUTCOME BASED EDUCATION

For the implementation of an outcome-based education, the first requirement is to develop an outcome-based curriculum and incorporate an outcome-based assessment in the education system. By going through outcome-based assessments evaluators will be able to evaluate whether the students have achieved the outlined standard, specific and measurable outcomes. With the proper incorporation of outcome-based education, there will be a definite commitment to achieve a minimum standard for all learners without giving up at any level. At the end of the program running with the aid of outcome-based education, a student will be able to arrive at the following outcomes:

PO1. **Engineering knowledge**: Apply the knowledge of mathematics, science, engineering fundamentals, and an engineering specialization to the solution of complex engineering problems.

PO2. **Problem analysis**: Identity, formulate, review research literature, and analyze complex engineering problems reaching substantiated conclusions using the first principles of mathematics, natural sciences, and engineering sciences.

PO3. **Design/development of solutions**: Design solutions for complex engineering problems and design system components or processes that meet the specified needs with appropriate consideration for public health and safety, and cultural, societal, and environmental considerations.

PO4. **Conduct investigations of complex problems**: Use research-based knowledge and research methods including design of experiments, analysis and interpretation of data, and synthesis of the information to provide valid conclusions.

PO5. **Modern tool usage**: Create, select, and apply appropriate techniques, resources, and modern engineering and IT tools including prediction and modeling to complex engineering activities with an understanding of the limitations.

PO6. **The engineer and society**: Apply reasoning informed by the contextual knowledge to assess societal, health, safety, legal and cultural issues and the consequent responsibilities relevant to the professional engineering practice.

PO7. **Environment and sustainability**: Understand the impact of professional engineering solutions in societal and environmental contexts, and demonstrate the knowledge of, and need for sustainable development.

PO8. **Ethics**: Apply ethical principles and commit to professional ethics and responsibilities and norms of the engineering practice.

PO9. **Individual and team work**: Function effectively as an individual, and as a member or leader in diverse teams, and in multidisciplinary settings.

PO10. **Communication**: Communicate effectively on complex engineering activities with the engineering community and with society at large, such as being able to comprehend and write effective reports and design documentation, make effective presentations, and give and receive clear instructions.

PO11. **Project management and finance**: Demonstrate knowledge and understanding of the engineering and management principles and apply these to one's own work, as a member and leader in a team, to manage projects and in multidisciplinary environments.

PO12. **Life-long learning**: Recognize the need for, and have the preparation and ability to engage in independent and life-long learning in the broadest context of technological change.

COURSE OUTCOMES

After completion of the course, the students will be able to:

CO-1: Apply basic ideas and principles of Electronics Engineering.

CO-2: To study of PN Junction devices and its applications

CO-3: To study the construction, working of a transistor and apply its characteristics

CO-4: To understand the concept of feedback amplifiers and oscillators

CO-5: To know the the fundamentals of Op-Amp its applications

CO-6: To understand the working principles of various measuring instruments and transducers

CO-7: To acquire the fundamental concepts of digital electronics.

Course Outcomes	Expected Mapping with Programme Outcomes *(1- Weak Correlation; 2- Medium correlation; 3- Strong Correlation)*											
	PO-1	**PO-2**	**PO-3**	**PO-4**	**PO-5**	**PO-6**	**PO-7**	**PO-8**	**PO-9**	**PO-10**	**PO-11**	**PO-12**
CO-1	3	3	2	1	-	1	-	-	-	-	-	1
CO-2	3	3	2	1	-	-	-	-	-	-	-	-
CO-3	3	3	2	1	-	-	-	-	-	-	-	-
CO-4	3	3	3	2	-	-	-	-	-	-	-	-
CO-5	3	3	3	1	-	-	-	-	-	-	-	-
CO-6	3	3	2	2	-	-	-	-	-	-	-	-
CO-7	3	3	3	1	-	1	-	-	-	-	-	-

GUIDELINES FOR TEACHERS

To implement Outcome Based Education (OBE) knowledge level and skill set of the students should be enhanced. Teachers should take major responsibility for the proper implementation of OBE. Some of the responsibilities (not limited to) for the teachers in the OBE system may be as follows:

- Within reasonable constraints, they should manoeuvre the time to the best advantage of all students.
- They should assess the students only upon certain defined criteria without considering any other potential ineligibility to discriminate against them.
- They should try to grow the learning abilities of the students to a certain level before they leave the institute.
- They should try to ensure that all the students are equipped with quality knowledge as well as competence after they finish their education.
- They should always encourage the students to develop their ultimate performance capabilities.
- They should facilitate and encourage group work and teamwork to consolidate newer approaches.
- They should follow Blooms taxonomy in every part of the assessment.

Bloom's Taxonomy

Level	Teacher should Check	Students should be able to	Possible Mode of Assessment
Create	Student's ability to create	Design or Create	Mini project
Evaluate	Student's ability to justify	Argue or Defend	Assignment
Analyse	Student's ability to distinguish	Differentiate or Distinguish	Project/Lab Methodology
Apply	Student's ability to use information	Operate or Demonstrate	Technical Presentation/ Demonstration
Understand	Student's ability to explain the ideas	Explain or Classify	Presentation/Seminar
Remember	Student's ability to recall (or remember)	Define or Recall	Quiz

GUIDELINES FOR STUDENTS

Students should take equal responsibility for implementing the OBE. Some of the responsibilities (not limited to) for the students in the OBE system are as follows:

- Students should be well aware of each UO before the start of a unit in every course.
- Students should be well aware of each CO before the start of the course.
- Students should be well aware of each PO before the start of the program.
- Students should think critically and reasonably with proper reflection and action.
- The learning of the students should be connected and integrated with practical and real-life consequences.
- Students should be well aware of their competency at every level of OBE.

ABBREVIATIONS AND SYMBOLS

List of Abbreviations

General Terms			
Abbreviations	**Full form**	**Abbreviations**	**Full form**
AC	Alternative Current	HWR	Half Wave Rectifier
ASIC	Application-Specific Integrated Circuits	JFET	Junction Field Effect Transistor
BCD	Binary Coded Decimal	LED	Light Emitting Diode
BJT	Bipolar Junction Transistor	LSB	Least Significant Bit
CB	Common Base Configuration	LVDT	Linear Variable Differential Transformer
CC	Common Collector Configuration	MOSFET	Metal Oxide Semiconductor Field Effect Transistor
CD	Diffusion Capacitance	MSB	Most Significant Bit
CE	Common Emitter Configuration	PIV	Peek Inverse Voltage
CMOS	Complementary Metal Oxide Semiconductor	PMMC	Permanent Magnet Moving Coil
CMRR	Common Mode Rejection Ratio	RAM	Random Access Memory
D- MOSFET	Depletion Metal Oxide Semiconductor Field Effect Transistor	RMS	Root Mean Square
DC	Direct Current	ROM	Read Only Memory
E- MOSFET	Enhancement Metal Oxide Semiconductor Field Effect Transistor	SCR	Silicon Controlled Rectifiers
EEPROM	Electrically Erasable Programmable Read Only Memory	SVRR	Supply Voltage Rejection Ratio
FET	Field Effect Transistor	VLSI	Very Large Scale Integration

List of Symbols

Symbols	Description	Symbols	Description
R_D	Static Resistance	μ	Amplification Factor
C_T	Transition Capacitance	I_{GSS}	Gate Cut-Off Current
C_D	Diffusion or storage Capacitance	R_{GS}	Input Resistance
η	The Ratio or Efficiency	g_m	Trans-Conductance
R_Z	Zener Resistance	A_f	Feedback Amplifier
V_Z	Zener Voltage	V_s	Signal Voltage
$P_Z$max	Maximum Power Dissipation	I_s	Signal Current
I_Z	Zener Current	I_{io}	Input Offset Current
I_C	Collector Current	I_B	Input Bias Current
I_E	Emitter Current	A	Large Signal Voltage Gain
S	Stability Factor	V_{CM}	Common Mode Voltage
S' & S''	Thermal Stability Factor	V_{oo}	Output Offset Voltage
I_{DSS}	Drain to Source Current	V_{io}	Input Offset Voltage
V_{GS}	Gate to Source Voltage	R_T	Resistance of the Thermistor
R_{DS}	Drain Source Resistance	G_f	Gauge Factors

LIST OF FIGURES

Unit 1 Diode and Its Application

Fig. 1.1 : P-N Junction with No bias condition 3

Fig. 1.2 : P-N Junction in reverse bias 4

Fig. 1.3 : P-N Junction in forward bias 4

Fig. 1.4 : The diode V-I characteristics 4

Fig. 1.5 : Basic structure of energy band diagram in semiconductor 5

Fig. 1.6 : Fermi level position in doped semiconductors 5

Fig. 1.7 : Band diagram before contact 5

Fig. 1.8 : Band diagram after contact(Ideal case) 6

Fig. 1.9 : (a) and (b) Intermediate steps of constructing the energy band diagram of a P-N junction, (c) and (d) The complete band diagram 6

Fig. 1.10 : Practical energy band diagram for a P-N junction 7

Fig. 1.11 : Energy band diagram under thermal equilibrium 8

Fig. 1.12 : Energy band diagram under external bias 8

Fig. 1.13 : P-N Junction in forward bias 9

Fig. 1.14 : P-N Junction with reverse bias 10

Fig. 1.15 : (a) A reverse-biased P-N junction, (b) band diagram without bias, and (c) energy band under reverse bias 11

Fig. 1.16 : (a) A forward-biased P-N junction, (b) band diagram without bias, and (c) energy band under forward bias 11

Fig. 1.17 : Depletion layer capacitance 12

Fig. 1.18 : Diode characteristics for different temperatures 14

Fig. 1.19 : A diode circuit 15

Fig. 1.20 : Switching timing of diode 16

Fig. 1.21 : Storage timing of diode 16

Fig. 1.22 : Block diagram of DC power supply 17

Fig. 1.23 : Half Wave Rectifier circuit and its output waveform 17

Fig. 1.24 (a) : Thevenin's equivalent of Half wave rectifier 20

Fig. 1.24 (b) : Variation in terminal voltage with load current for ideal & practical power supply 20

Fig. 1.25 : A Centertap Full Wave Rectifier Circuit 22

Fig. 1.26 : Full Wave Bridge Rectifier Circuit 25

Fig. 1.26(a) : Inductor filter circuit 27

Fig. 1.26(b) : Output of inductor filter 27

Fig. 1.27 : Output waveform of an inductor filter 29

Fig. 1.28 : Capacitor filter circuit 29

Fig. 1.29 : Output of capacitor filter 30

Fig. 1.30 : Zener diode as a voltage regulator 31

Fig. 1.31 : Band diagram of Zener breakdown 33

Fig. 1.32 : I-V characteristics of Zener and Avalanche Breakdown 34

Fig. 1.33 : A practical diode under DC operating conditions 35

Fig. 1.34(a) : Forward biased DC diode model 35

Fig. 1.34(b) : Reverse biased DC diode model 35

Fig. 1.35 : Reverse biased ac diode model 36

Fig. 1.36 : Forward biased ac diode model 36

Fig. 1.37 : Diode as a switch 37

Fig. 1.38 : Symbol and circuit diagram of a LED 37

Fig. 1.39 : Symbol and I-V Characteristics of a Zener diode 38

Fig. 1.40 : The details of a photo diode 39

Fig. 1.41 : V-I Characteristics of a photo diode 39

Fig. 1.42 : (a) SCR symbol; (b) basic construction 40

Fig. 1.43 : Two transistor terminology of a SCR 41

Fig. 1.44 : "Off" state of the SCR 41

Fig. 1.45 : "On" state of the SCR 42

Fig. 1.46 : SCR characteristics 42

Unit 2 Transistor Characteristics

Fig. 2.1 : Construction of BJT-PNP & NPN 53

Fig. 2.2 : Depletion Region and Barrier Potential 55

Fig. 2.3 : Transistor Operation (NPN) 56

Fig. 2.4 : Transistor Operation (PNP) 57

Fig. 2.5 : Common Base Configuration 57

Fig. 2.6 : I_{CBO} In CB Configuration 58

Fig. 2.7 : Common Emitter Configuration 58

Fig. 2.8 : Common Collector Configuration 59

Fig. 2.9 : Circuit under consideration 60

Fig. 2.10 : DC Load Line 60

Fig. 2.11 : Potential divider bias 62

Fig. 2.12 : Self bias circuit 64

Fig. 2.13 : Thermistor bias Compensation 64

Fig. 2.14 : Sensistor bias Compensation 65

Fig. 2.15 : Transistor as Amplifier 65

Fig. 2.16 : Construction and symbol of an (a) n- channel and (b) p-channel JFET 67

Fig. 2.17 : N-Channel FET with applied drain voltage polarities 67

Fig. 2.18 : The rise in depletion width with applied $V_{DS}(V_{DD})$ 68

Fig. 2.19 : Depletion Region development due to internal voltage drop 68

Fig. 2.20 : Pinch-off occurs at $V_{GS} = V_P$ 70

Fig. 2.21 : Output Characteristics 70

Fig. 2.22 : Construction and symbol of a D-MOSFET 72

Fig. 2.23 : Operation of a D-MOSFET at V_{GS} =0 72

Fig. 2.24 : Operation of a D-MOSFET at V_{GS} =negative 73

Fig. 2.25 : Operation of a D-MOSFET at V_{GS} =positive 73

Fig. 2.26 : The drain –source characteristics and transfer characteristics 74

Fig. 2.27 : Symbol of a p-channel D-MOSFET 75

Fig. 2.28 : Construction and symbol of an E-MOSFET 75

Fig. 2.29 : Drain Source Characteristics 76

Fig. 2.30 : Basic CMOS configuration 77

Fig. 2.31 : Symbol of an NMOS and PMOS transistor 77

Fig. 2.32 : CMOS using Pull up and Pull Down 78

Fig. 2.33 : CMOS Inverter 79

Unit 3 Feedback Amplifier and Oscillators

Fig. 3.1 : A single-loop feedback system 91

Fig. 3.2 : Mixer Networks: (a) Voltage Comparator (b) Current Comparator 92

Fig. 3.3 : Sampling Networks: (a) Voltage Sampling (b) Current Sampling 92

Fig. 3.4 : Feedback amplifiers connection type: (a) Voltage series feedback (b) Current series feedback (c) Current-shunt feedback d) Voltage-shunt feedback 93

Fig. 3.5 : Voltage series feedback connection 95

Fig. 3.6 : Voltage-shunt feedback connection 95

Fig. 3.7 : Current-Series feedback connection 97

Fig. 3.8 : Effect of negative feedback on gain and bandwidth 98

Fig. 3.9 : Difference between amplifier and oscillator 99

Fig. 3.10 : Tank circuit (LC-tuned circuit) 99

Fig. 3.11 : Charging of an Inductor 99

Fig. 3.12 : Discharging of an Inductor 100

Fig. 3.13 : Damped Sinusoidal waveform 100

Fig. 3.14 : Block diagram of positive feedback amplifier 101

Fig. 3.15 : Oscillator classification chart 103

Fig. 3.16 : RC phase shift oscillator 104

Fig. 3.17 : A Wien Bridge Oscillator 105

Fig. 3.18 : Circuit diagram of Hartley oscillator 106

Fig. 3.19 : Circuit diagram of Colpitts oscillator 107

Fig. 3.20 : Electrical equivalent circuit of crystal 108

Fig. 3.21 : Transistorized crystal oscillator 108

Fig. 3.22 : Wien Bridge oscillator circuit 110

Unit 4 Operational Amplifier and Application

Fig. 4.1 : Block diagram of Op-Amp 116

Fig. 4.2 : Symbol of Op-Amp 117

Fig. 4.3 : Pin configuration of IC 741 Op-Amp 117

Fig. 4.4 : Op-Amp equivalent circuit 118

Fig. 4.5 : Op-Amp equivalent circuit for an ideal case 118

Fig. 4.6 : Circuit arrangement for calculation of offset voltage 119

Fig. 4.7 : Op-Amp inverting amplifier 120

Fig. 4.8 : Open loop inverting amplifier 121

Fig. 4.9 : Open-loop non-inverting amplifier 122

Fig. 4.10 : Differential amplifier using OP-Amp 122

Fig. 4.11 : Non-Inverting Amplifier with Feedback 123

Fig. 4.12 : Equivalent circuit for input impedance calculation 124

Fig. 4.13 : Equivalent circuit for output impedance calculation 125

Fig. 4.14 : Inverting Amplifier with feedback 126

Fig. 4.15 : Voltage follower circuit 126

Fig. 4.16 : Op-Amp as an Inverter 127

Fig. 4.17 : Op-Amp as Current to voltage converter 127

Fig. 4.18 : Inverting summing amplifier 128

Fig. 4.19 : Non-inverting summing amplifier 129

Fig. 4.20 : Op-Amp as a Subtractor 131

Fig. 4.21 : Op-Amp Integrator circuit 132

Fig. 4.22 : Op-Amp differentiator circuit 133

Fig. 4.23 : Input and Output waveform of a Differentiator 133

Fig. 4.24 : Ideal filter response 134

Fig. 4.25 : (a) A first-order low pass filter (b) Filter response 134

Fig. 4.26 : (a) Second order Low pass filter (b) Filter response 135

Fig. 4.27 : (a) First order High pass filter (b) Filter response 135

Fig. 4.28 : (a) Second order High pass filter (b) Filter response 136

Fig. 4.29 : A bandpass active filter 136

Unit 5 Measuring Instruments & Transducers

Fig. 5.1 : Construction of D' Arsonval Instrument 159

Fig. 5.2 : Instrument for Permanent Magnet Moving Coil (PMMC) 160

Fig. 5.3 : Ohmmeter 162

Fig. 5.4 : A series ohmmeter 163

Fig. 5.5 : Scale of a series ohmmeter 163

Fig. 5.6 : (a) Shunt type ohmmeter (b) Scale of a shunt ohmmeter 164

Fig. 5.7 : Multirange Ohmmeter 165

Fig. 5.8 : Galvanometer 165

Fig. 5.9 : Concept of a potentiometer 167

Fig. 5.10 : Potentiometer 168

Fig. 5.11 : A rotary potentiometer 168

Fig. 5.12 : Construction of a moving iron frequency meter 169

Fig. 5.13 : Construction of an Electrodynamic frequency meter 170

Fig. 5.14 : Types of Thermistors 179

Fig. 5.15 : Resistance-Temperature characteristics 179

Fig. 5.16 :The construction of LVDT 181

Fig. 5.17 : Hall Effect element 184

Fig. 5.18 : Parallel plate capacitive transducer 186

Fig. 5.19 : Variation in the distance between two plates 187

Fig. 5.20 : Capacitance between two plates 188

Fig. 5.21 : Piezoelectric transducer 190

Fig. 5.22 : Construction of a seismic transducer 192

Unit 6 Introduction to Digital Electronics

Fig. 6.1 : Illustration of 9's complement arithmetic 206

Fig. 6.2 : Illustration of 10's complement arithmetic 207

Fig. 6.3 : A Two-input OR gate and its truth table 233

Fig. 6.4 : A Two-input AND gate and its truth table 234

Fig. 6.5 : Symbol of a NOT gate and its truth table 234

Fig. 6.6 : A Two-input Ex-OR gate 235

Fig. 6.7 : A Two-input NAND gate and its truth table 235

Fig. 6.8 : A Two-input NOR gate and its truth table 236

Fig. 6.9 : A Two-input Ex-NOR gate and its truth table 236

Fig. 6.10 : NAND gate as a Universal gate 237

Fig. 6.11 : NOR gate as a Universal gate 237

Fig. 6.12 : A combinational logic 240

Fig. 6.13 : Classification of Combination Logic 241

Fig. 6.14 : Sequential logic 242

LIST OF TABLES

Unit 1 Diode and Its Application

Table. 1.1 : Comparison of Zener and Avalanche Breakdown 34

Unit 3 Feedback Amplifier and Oscillators

Table. 3.1 : Summary of Gain, Feedback, and Gain with feedback from Fig.3.4 94

Table. 3.2 : Summarizes the effect of feedback on input and output impedance. 97

Unit 5 Measuring Instruments & Transducers

Table. 5.1 :Key difference between a sensor and a transducer 172

Table. 5.2 : Gauge Factor 177

Unit 6 Introduction to Digital Electronics

Table. 6.1 : Octal to Binary Conversion 215

Table. 6.2 : Hexadecimal to Binary Conversion 222

Table. 6.3 : BCD code representations 224

Table. 6.4 : Excess-3 code for the decimal numbers 226

Table. 6.5 : Gray code 227

Table. 6.6 : Comparison between the Combinational and Sequential Logic Circuits 243

CONTENTS

Foreword iii
Acknowledgment iv
Preface v
Outcome Based Education vi
Course Outcomes vii
Guidelines for Teachers viii
Guidelines for Students viii
Abbreviations and Symbols ix
List of Figures xi
List of Tables xix

Unit 1: Diode and Its Applications **1-50**
Rationale 1
Unit Outcomes 1
Learning Objective 1
Mapping the unit outcome with course outcomes 2
1.1 *Review of a P-N junction* 3
1.2 *The V-I Characteristics* 3
1.3 *Semiconductor band and Fermi energy levels* 5
1.3.1 Position of Fermi level in doped Semiconductors 5
1.3.2 Energy Band Diagram for a P-N junction 5
1.3.3 Built-in voltage of the P-N junction 6
1.4 *P-N Junction with external bias* 7
1.4.1 Zero bias voltage 7
1.4.2 Forward bias condition 8
1.4.3 Energy band diagrams for P-N junction with external bias 11
1.5 *Static and dynamic resistance of a P-N junction diode* 11
1.6 *Diode capacitance* 12
1.7 *Temperature Effect on Diode* 14
1.8 *Switching time of a Diode* 15
1.9 *DC Power Supply* 16
Solved Examples 43
Exercises Questions 45
Self-Study Questions 45
Multiple Choice Questions 46
Attainment & Gap Analysis 50

Unit 2: Transistor Characteristics **51-88**
Rationale 50
Unit Outcomes 50

	Learning Objective	*50*
	Mapping the unit outcome with the course outcomes	*50*
2.1	*Bipolar Junction Transistor (BJT)*	*53*
2.2	*Modes (Regions) of operation of BJT*	*54*
2.3	*Transistor operation in linear mode*	*55*
	2.3.1 For NPN transistor	*55*
	2.3.2 For PNP transistor	*56*
2.4	*Transistor Configuration*	*57*
	2.4.1 Common Base Configuration (CB)	*57*
	2.4.2 Common Emitter Configuration(CE)	*58*
	2.4.2 Common Collector Configuration(CC)	*59*
2.5	*DC Load Line*	*59*
2.6	*Transistor biasing*	*61*
2.7	*Thermal Stability*	*63*
	2.7.1 Generalized Expression for S	*63*
	2.7.2 Bias Compensation	*64*
2.8	*Transistor as Amplifier*	*65*
2.9	*The Field Effect Transistor (FET)*	*66*
	2.9.1 Junction Field effect transistors (JFET)	*66*
	2.9.2 Operation of a JFET (n-channel)	*67*
	2.9.3 FET Parameters	*70*
	2.9.4 Comparison between BJT and FET	*71*
2.10	*MOSFET (Metal oxide semiconductor field effect transistor)*	*71*
	2.10.1 D-MOSFET (n-channel)	*72*
	2.10.2 E-MOSFET (n-channel)	*75*
2.11	*CMOS*	*77*
	Solved Example	*80*
	Exercises Questions	*83*
	Self-Study Questions	*84*
	Multiple Choice Questions	*84*
	Attainment & Gap Analysis	*88*

Unit 3: Feedback Amplifier and Oscillators ***89-113***

	Rationale	*89*
	Unit Outcomes	*89*
	Learning Objective	*89*
	Mapping the unit outcome with the course outcomes	*89*
3.1	*The concept of feedback*	*91*
3.2	*Feedback connection types*	*93*
3.3	*The gain with feedback*	*94*
3.4	*Input Impedance with feedback*	*95*
3.5	*Output Impedance with feedback*	*96*
3.6	*General characteristics of a negative feedback amplifier*	*97*
3.7	*Oscillators*	*99*
3.8	*Block Diagram of Positive Feedback Amplifier*	*101*
3.9	*Classification of Oscillator*	*102*

3.10 RC phase shift oscillator 103
3.11 Wien Bridge Oscillator 105
3.12 LC Oscillator 106
3.12.1 Hartley Oscillator 106
3.12.2 Colpitts Oscillator 107
3.12.3 Crystal Oscillator 107
Solved Example 109
Exercises Questions 110
Self-Study Questions 110
Multiple Choice Questions 111
Attainment & Gap Analysis 113

Unit 4: Operational Amplifier and Applications ***114-155***
Rationale 114
Unit Outcomes 114
Learning Objective 114
Mapping the unit outcome with the course outcomes 114
4.1 General characteristics of an Op-Amp 116
4.2 Important Specification of Op-Amp 118
4.3 Concept of Virtual ground 120
4.4 Open Loop configuration of an Op-Amp 121
4.4.1 Inverting Amplifier 121
4.4.2 Non-Inverting Amplifier 121
4.4.3 Differential Amplifier 122
4.5 Close loop Configurations of an Op-Amp 122
4.5.1 Non-Inverting Amplifier with Feedback 123
4.5.2 Inverting Amplifier with Feedback 125
4.5.3 Buffer Amplifier or Voltage Follower 126
4.5.4 Inverter (Sign Changer) 127
4.5.5 Current to voltage converter 127
4.6 Application of Op-Amp 128
4.6.1 Summing, Scaling, and Averaging Amplifiers 128
4.6.2 OP-Amp as an Subtractor 130
4.6.3 OP-Amp as an Integrator 131
4.6.4 OP-Amp as Differentiator 132
4.7 Active Filter using Op-Amp 134
4.7.1 Low pass filter 134
4.7.2 High pass Active filter 135
4.7.3 High pass Active filter 136
Solved Example 136
Exercises Questions 149
Self-Study Questions 151
Multiple Choice Questions 151
Attainment & Gap Analysis 155

Unit 5: Rotational Motion **156-201**

Rationale 156
Unit Outcomes 156
Learning Objective 156
Mapping the unit outcome with the course outcomes 156
5.1 *D' Arsonval Movement* 159
5.2 *Permanent Magnet Moving Coil Instrument* 160
5.3 *Ohmmeter* 162
5.3.1 Series Ohmmeter 163
5.3.2 Shunt Ohmmeter 164
5.3.3 Multi range Ohmmeter 164
5.4 *Galvanometer* 165
5.5 *Potentiometer* 167
5.6 *Frequency Meters* 169
5.6.1 Moving Iron Frequency Meter 169
5.6.2 Electrodynamic Frequency Meter 170
5.6.3 Vibrating-reed Frequency Meter 171
5.7 *Sensors and Transducers* 170
5.7.1 Electrical Transducers 172
5.7.2 Characteristics of a Transducers 173
5.7.3 Factor influencing the choice of transducers 173
5.7.4 Classification of Transducer 174
5.8 *Strain Gauges* 175
5.9 *Thermistors* 177
5.10 *Liner variable differential transformer (LVDT)* 180
5.11 *Hall Effect Transducer* 183
5.12 *Capacitive Transducer* 185
5.13 *Piezoelectric Transducer* 189
5.14 *Seismic Transducer* 191
Solved Example 193
Exercises Questions 196
Multiple Choice Questions 197
Attainment & Gap Analysis 201

Unit 6: Introduction to Digital Electronics **202-253**

Rationale 202
Unit Outcomes 202
Learning Objective 202
Mapping the unit outcome with the course outcomes 203
6.1 *The number system* 204
6.1.1 Decimal Number System 205
6.1.2 Binary Number System 208
6.1.3 Octal Number System 208
6.1.4 Hexadecimal Number System 208
6.2 *Number Representation in Binary System* 209
6.2.1 Sign-Bit Magnitude 209
209

6.3 The Decimal Equivalent 211
6.3.1 Binary-to-Decimal Conversions 211
6.3.2 Octal-to-Decimal Conversions 211
6.3.3 Hexadecimal-to-Decimal Conversions 211
6.3.4 Decimal-to-Binary Conversions 211
6.3.5 Decimal-to-Octal Conversions 212
6.3.6 Decimal-to-Hexadecimal Conversions 213
6.3.7 Binary-Octal and Octal-Binary Conversions 214
6.3.8 Binary-Hex and Hex-Binary Conversions 218
6.3.9 Hex-Octal and Octal-Hex Conversions 223
6.4 Binary Codes 224
6.4.1 Binary Coded Decimal (BCD) 224
6.4.2 Excess-3 Code 226
6.4.3 Gray Code 227
6.5 The Digital Arithmetic 230
6.6 The Logic Gates 233
6.6.1 The OR Gate 233
6.6.2 The AND Gate 234
6.6.3 The NOT Gate 234
6.6.4 The EXCLUSIVE-OR Gate 234
6.6.5 The NAND Gate 235
6.6.6 The NOR Gate 236
6.6.7 The EXCLUSIVE-NOR Gate 236
6.6.8 The Universal Gates 236
6.7 The Boolean algebra 236
6.8 The Digital Circuits 237
6.8.1 The Combinational Circuits 239
6.8.2 The Sequential Circuits 239
Solved Example 241
Exercises Questions 244
Self-Study Questions 248
Multiple Choice Questions 249
Attainment & Gap Analysis 250
253

References for Further Learning 255
CO and PO Attainment Table 256
Index 257

1. Diode and Its Applications

RATIONALE

It is now more than 80 years since the first P-N junction was introduced. Since then, it becomes an integral part of the majority of IC designs.

In this chapter, we will study the details of a semiconductor diode, the construction and working principles of a few advanced and application-oriented diodes.

UNIT OUTCOMES

U1-O1: Unit-1 Learning Outcome-1

To know about the basic operation of a P-N junction diode, its breakdown mechanism, equivalent circuits, and loadline analysis.

U1-O2: Unit-1 Learning Outcome-2

To know about the construction and operation of different types of rectifiers and different breakdown mechanisms of the P-N junction diodes.

U1-O3: Unit-1 Learning Outcome-3

To know the operation and applications of a Zener diode, opto-electronic devices such as LEDs and photo-diode, and SCR.

LEARNING OBJECTIVES

LO1: Study of fundamentals of a P-N junction diode; ideal versus practical.

LO2: Importance of Diode equivalent circuits, their resistance levels, and load line analysis.

LO3: Working of a diode as a switch.

LO4: Study of construction and operation of half-wave and full-wave rectifiers; with and without filters.

LO5: Importance of breakdown mechanism.

LO6: operation and applications of a Zener diode, various opto-electronic devices such as LEDs and photo-diode, SCR.

MAPPING THE UNIT OUTCOMES WITH THE COURSE OUTCOMES

Unit Outcome	EXPECTED MAPPING WITH COURSE OUTCOMES (1- Weak Correlation; 2- Medium correlation; 3- Strong Correlation)						
	CO-1	**CO-2**	**CO-3**	**CO-4**	**CO-5**	**CO-6**	**CO-7**
U1-O1	**3**	**3**	--	--	--	--	--
U1-O2	**3**	**3**	--	--	--	--	--
U1-O3	**3**	**3**	--	--	--	--	--

Interesting Facts:

1. **In 1939, Russell Ohl** of **Bell Laboratories** discovers the P-N junction and photovoltaic effects in silicon that lead to the development of junction transistors and solar cells.
2. A semiconductor is made using materials such as **silicon, germanium, and gallium arsenide.** These materials are also used to make P-N junction diodes. Silicon is used over germanium when creating these junction diodes.
3. **Photodiodes** are special P-N junction diodes operated in reverse bias. They are mainly designed for detecting optical signals.

Video Resources:

Sr	Title	URL	QR Code
1.	Semiconductor: P-Type and N-Type, Intrinsic and Extrinsic	https://www.youtube.com/watch?v=6dUpV2ovqog	
2.	PN Junction Diode and V-I Characteristics	https://www.youtube.com/watch?v=o-Rya9KZYY4	
3.	Rectifier Operation	https://www.youtube.com/watch?v=5cbQNfO0Mwg	
4.	What is a Zener Diode	https://www.youtube.com/watch?v=XhQqtdTlRus	
5.	What is Photodiode	https://www.youtube.com/watch?v=8k9UIlwo7W4	
6.	Light Emitting Diode (LED) Working Principle	https://www.youtube.com/watch?v=wl45Rrt4j2U	
7.	SCR	https://www.youtube.com/watch?v=RJ43fnX-LsM	

1.1 Review of a *P-N* junction

There are two types of extrinsic semiconductors. Extrinsic semiconductors with a more significant electron concentration than hole concentration are known as N-type semiconductors. The phrase 'N-type' comes from the negative charge of the electron. In N-type semiconductors, electrons are the majority carriers, and holes are the minority carriers. N-type semiconductors are created by doping an intrinsic semiconductor with donor impurities.

In an N-type semiconductor, the Fermi energy level is greater than that of the intrinsic semiconductor and lies closer to the conduction band than the valence band. Arsenic has five valence electrons; however, only 4 of them form part of covalent bonds. The 5th electron is then free to take part in conduction. The electrons are said to be the majority carriers, and the holes are the minority carriers.

The P-type semiconductors have a more significant hole concentration than electron concentration. The phrase 'P-type' refers to the positive charge of the hole. In P-type semiconductors, holes are the majority carriers, and electrons are the minority carriers. P-type semiconductors are created by doping an intrinsic semiconductor with acceptor impurities. P-type semiconductors have Fermi energy levels below the intrinsic Fermi energy level. The Fermi energy level lies closer to the valence band than the conduction band in a P-type semiconductor. For example, gallium has three valence electrons; however, there are four covalent bonds to fill. The 4th bond, therefore, remains vacant, producing a hole. The holes are said to be the majority carriers, and the electrons are the minority carriers.

When P–type, and N–type materials come in close contact, electrons start diffusing from N-type material into P-material. At the same time, holes also begin to diffuse from P-type material into N-material. Every electron transfers a negative charge (-q) onto the P-side and leaves an uncompensated (+q) charge of the donor on the N-side. Every hole creates one positive charge (q) on the N-side and (-q) on the P-side.

A negative charge prevents electrons from further diffusion, and a Positive charge stops holes from further diffusion. The diffusion forms a dipole charge layer at the P-N junction interface. It is called a potential barrier, as shown in Fig 1.1. There is a "built-in" voltage at the P-N junction interface that prevents penetration of electrons into the P-side and holes into the N-side.

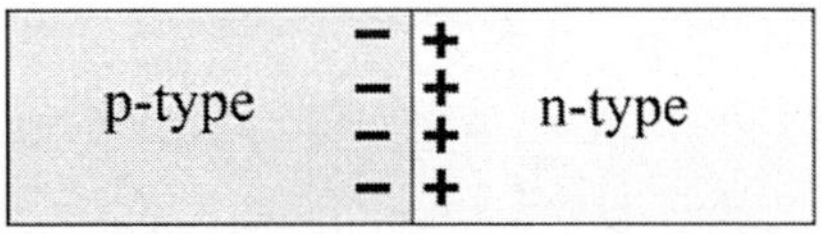

Fig. 1.1: P-N Junction with no bias condition

1.2 The V-I Characteristics

The polarity in Fig. 1.2 attracts the holes to the left and the electrons to the right. According to the current continuity law, the current can only flow if all the charged particles form a closed loop.

However, there are very few holes in the N-type material and few electrons in the P-type material. Therefore, very few carriers are available to support the current through the junction plane. For the voltage polarity shown, the current is nearly zero.

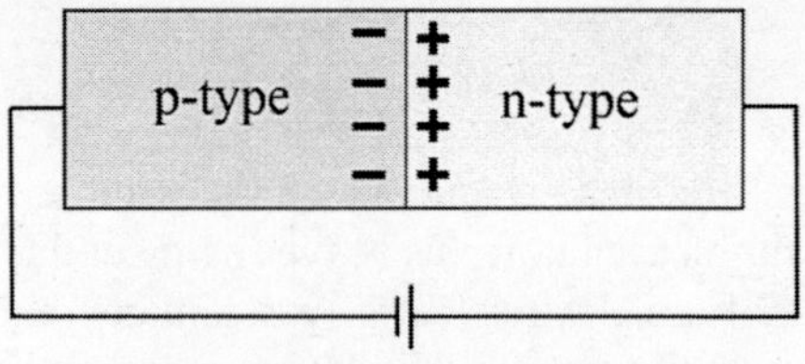

Fig. 1.2: P-N Junction in reverse bias

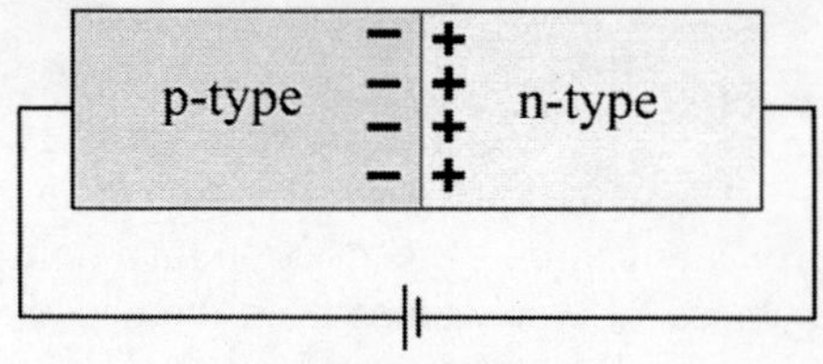

Fig.1.3: P-N Junction in forward bias

The polarity shown in Fig. 1.3 attracts electrons to the left and holes to the right. There are plenty of electrons in the N-type material and plenty of holes in the P-type material. Therefore, there are a lot of carriers available to cross the junction. When the voltage applied is lower than the built-in voltage, the current is approximately zero. When the voltage exceeds the built-in voltage, the current can flow through the P-N junction.

The semiconductor diode consists of a P-N junction with two contacts attached to the ***p*** and ***n*** sides. The complete current equation of the P-N junction diode is given below-

$$I = I_S\left[\exp\left(\frac{qV}{kT}\right) - 1\right]$$

k is Boltzmann constant

T is the temperature in K

At room temperature($T \approx 300K$), $\left(\frac{kT}{q}\right) \approx 0.026\ V$

I_S is a saturation current, typically I_S is in the range of $I_S \approx 10^{-17}$ to $10^{-11}\ A$

When the voltage V is negative ("reverse" polarity), the exponential term ≈ -1; The diode current is $\approx I_S$ (very small).

When the voltage V is positive ("forward" polarity), the exponential term increases rapidly with V, and the current is high.

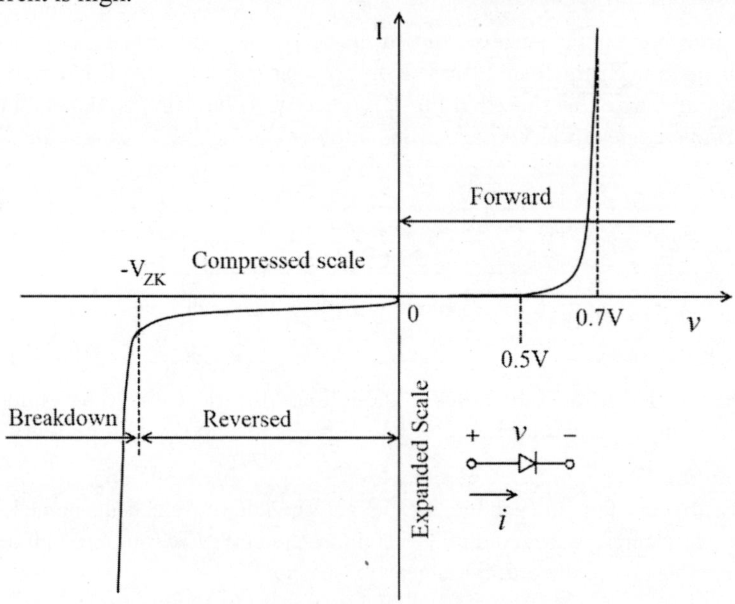

Fig. 1.4: The diode V-I characteristics

1.3. Semiconductor band and Fermi energy levels

Fermi energy E_F is the average energy of all the free carriers in a sample. In equilibrium, the Fermi energy must be uniform over the semiconductor (compared to the temperature distribution over any sample in equilibrium). Fig. 1.5 shows the basic structure of the energy band diagram in a semiconductor.

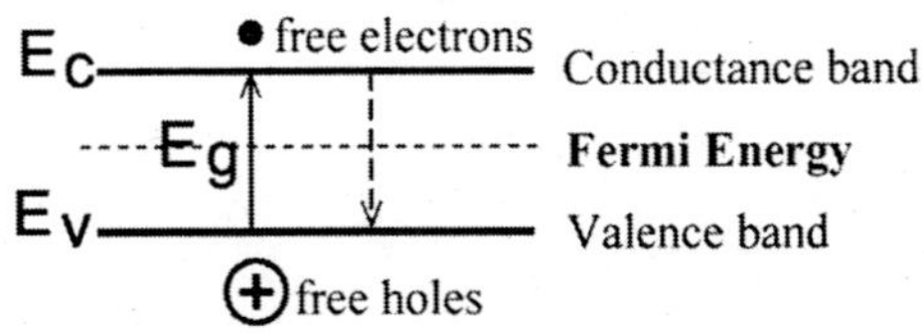

Fig. 1.5: Basic structure of energy band diagram in semiconductor

1.3.1. Position of Fermi level in doped semiconductors

In an intrinsic semiconductor, the Fermi level lies precisely at the center of the band gap. While in the extrinsic semiconductor Fermi level will not be in the center. It is close to the conduction band in N-type and the valance band in P-type. Fig. 1.6 shows the Fermi level position in doped semiconductors.

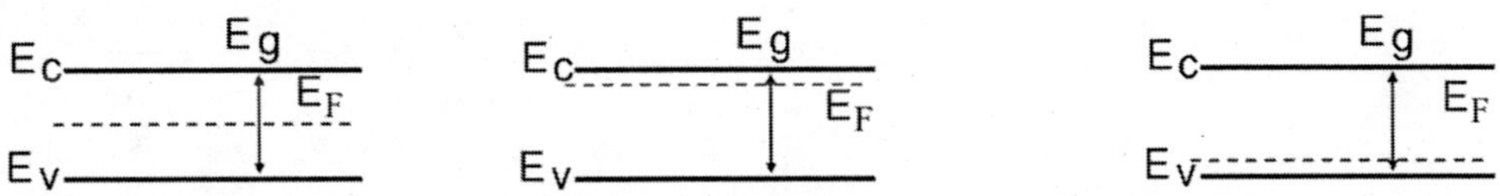

Intrinsic Semiconductor | Donor-doped semiconductor (N-type) | Acceptor doped semiconductor (P-type)

$n = p = n_i$ | $n'' p$ | $p'' n$

$E_F \sim (E_C + E_V)/2$ | $E_F \sim E_C$ | $E_F \sim E_V$

Fig.1.6: Fermi level position in doped semiconductors

1.3.2. Energy Band Diagram for a P-N junction

The energy band diagram for the junction can be understood as shown in Fig. 1.7.

1. Two separate bits of semiconductor first is N-type, the other is P-type

2. When bits are joined together: not in equilibrium $E_F(n) \neq E_F(p)$

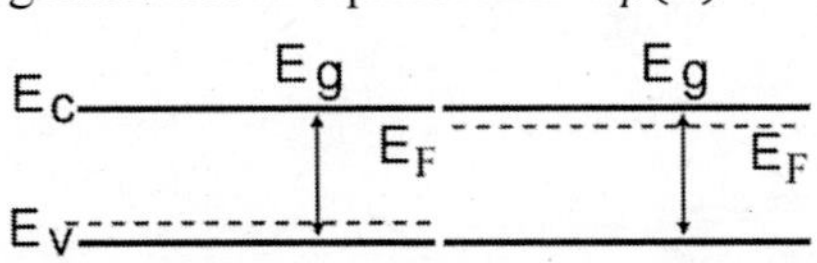

Fig. 1.7: Band diagram before contact

3. This state is unstable regarding its energy (once n and p are in contact). On the N-side, there are many electrons in the conduction band, and on the P-side, there are many holes in the valence band. Thus, due to the concentration gradient, charge carriers will flow from higher to lower concentrations. One needs to consider the flow of both electrons and holes.

Electrons close to the interface flow from ***n*** to the ***p*** side, and holes close to the interface flow from ***p*** to the ***n*** side. This flow continues until Fermi levels equilibrate.

The balance is achieved by electrons diffusing into a P-side (bringing an extra negative charge) and by the holes diffusing into an N-side (carrying an additional negative charge). These recombinations grow a layer on both sides of the barrier depleted of free carriers. Thus, the band diagram after contact, i.e., after the achievement of the equilibrium, is shown in the figure.

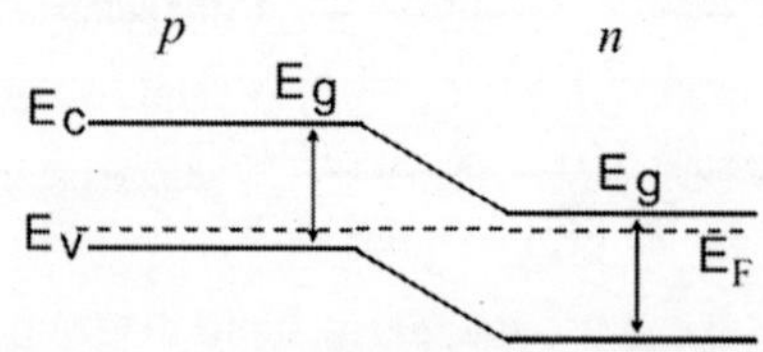

Fig. 1.8: Band diagram after contact (Ideal case)

Fig. 1.9 depicts the energy band diagram for the P-N junction.

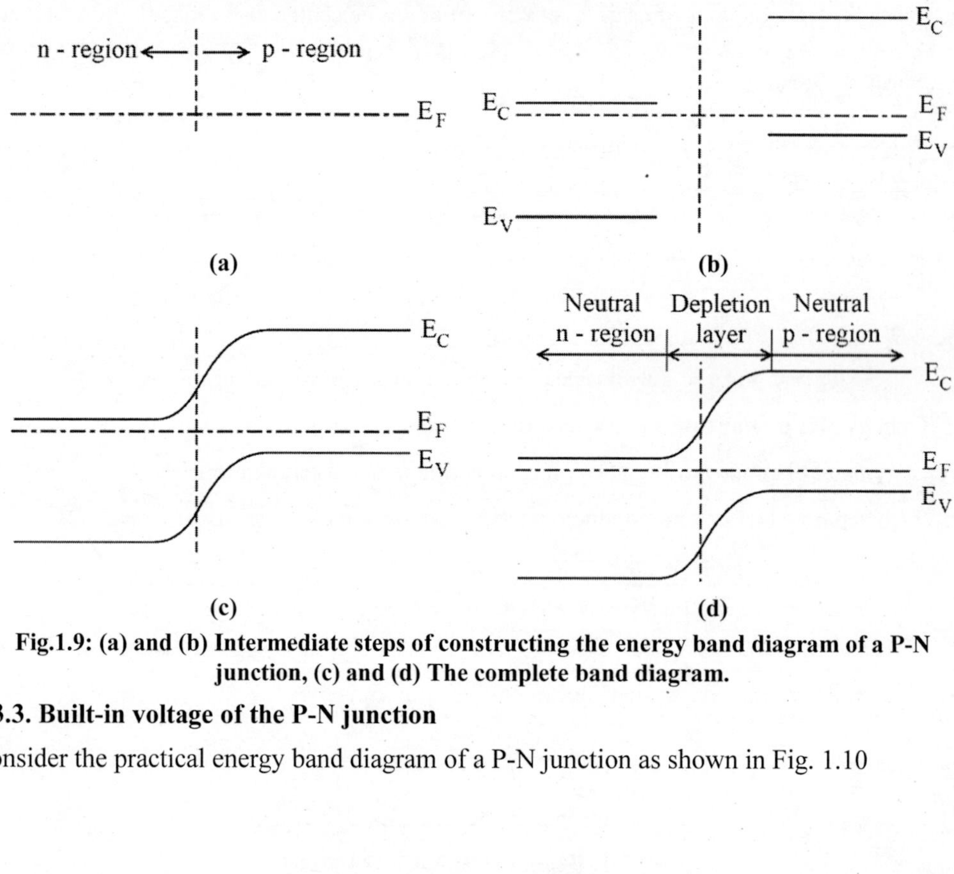

Fig.1.9: (a) and (b) Intermediate steps of constructing the energy band diagram of a P-N junction, (c) and (d) The complete band diagram.

1.3.3. Built-in voltage of the P-N junction

Consider the practical energy band diagram of a P-N junction as shown in Fig. 1.10

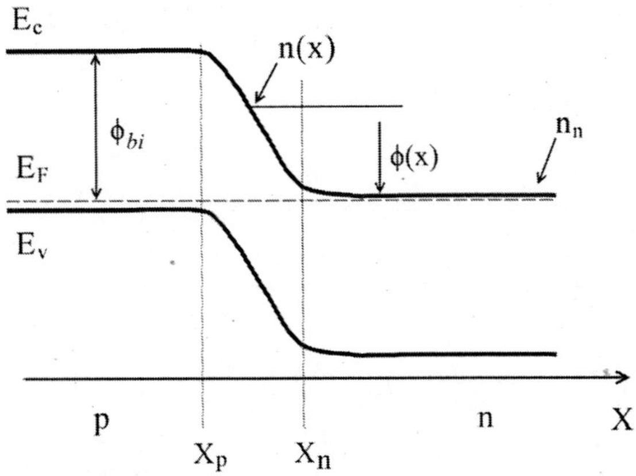

Fig. 1.10: Practical energy band diagram for a P-N junction

Far from the junction, on the N-side, $n = N_D$, at any arbitrary point x inside the transition region (between x_n and x_p);

$$n(x) = N_D e^{-\frac{\phi(x)}{kT}}$$

where $\phi(x)$ is the potential barrier at the point x

At the point located far in the p-region, the potential barrier flattens out and reaches ϕ_{bi}; at this point:

$$n_p = N_D e^{-\frac{\phi_{bi}}{kT}}$$

On the other hand, for the point in the p-material:

$$n_p = \frac{n_i^2}{N_A} = N_D e^{-\frac{\phi_{bi}}{kT}}$$

From this, the built potential barrier, i.e., the energy barrier, in joules or eV

$$\phi = kT.\ln\left(\frac{N_D N_A}{n_i^2}\right)$$

The voltage corresponding to the energy barrier

$$V_{bi} = \frac{\phi_{bi}}{q}$$

$$\Rightarrow V_{bi} = \frac{kT}{q}.\ln\left(\frac{N_D N_A}{n_i^2}\right)$$

1.4. P-N Junction with external bias

To understand the behavior of a P-N junction under external bias, two conditions, namely working under zero applied voltage and working with the application of the external bias, need to be considered.

1.4.1. Zero bias voltage

Fig. 1.11 shows the band diagram under thermal equilibrium for a P-N junction.

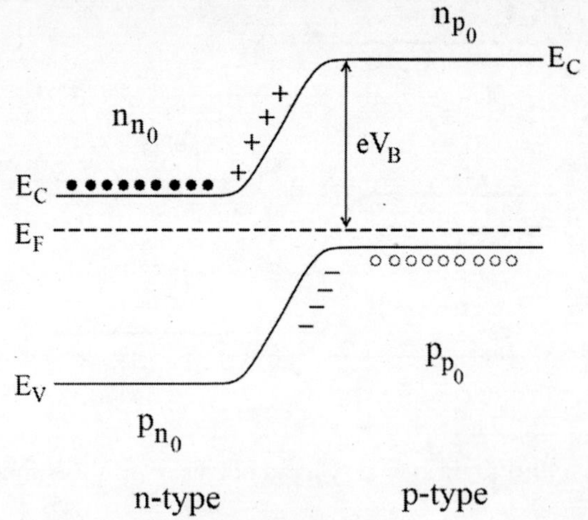

Fig. 1.11: Energy band diagram under thermal equilibrium

We know the ***n***- side has electrons as the majority of charge carriers, and the *P-type* has holes as the majority of charge carriers. Thus, away from the depletion region, the *N-side* supplies free electrons (n_{n_0} large) (majority) and *P-side* supplies free holes (p_{p_0} large) (majority).

As the holes are minority charge carriers on the *N-side* and electrons are minority charge carriers on the *P-side*, we get a small concentration n_{p_0} of free electrons on the *P-side* (minority) and a small concentration p_{n_0} of free holes on the *N-side* (minority).

Potential drop V_B across the junction (with zero applied bias) can be given as-

$$\frac{n_{p_0}}{n_{n_0}} = \frac{p_{n_0}}{p_{p_0}} = \exp\left(-\frac{eV_B}{kT}\right)$$

1.4.2. Forward bias condition

When the external bias voltage V_{ext} is applied, equilibrium is disturbed, as shown in Fig. 1.12.

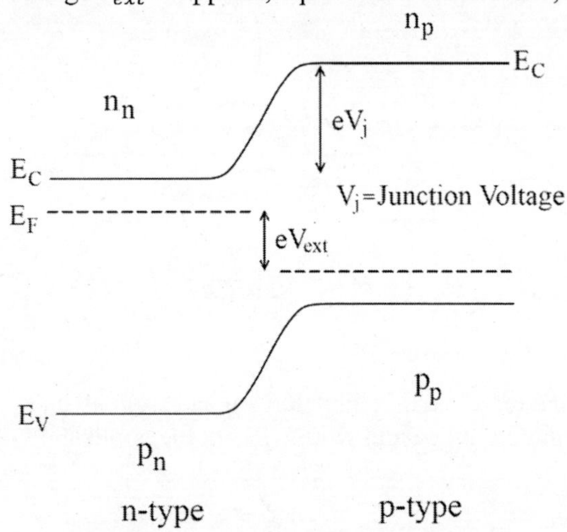

Fig. 1.12 Energy band diagram under external bias

1.4.2.1. Forward Bias:

Electrons in the P-type material, near the positive terminal of the supply, break their electron pair bonds and enter the supply, thereby producing new holes. Electrons from the supply's negative terminal enter the N-type material and migrate toward the junction.

Free electrons from the N-type material flow across the junction and move into the holes migrated from the positive terminal. This current flow will continue as long as the external bias is connected, known as the forward current flow. In forward bias, the potential drop across the junction is reduced. Fig.1.13 shows the forward bias condition of the P-N junction diode.

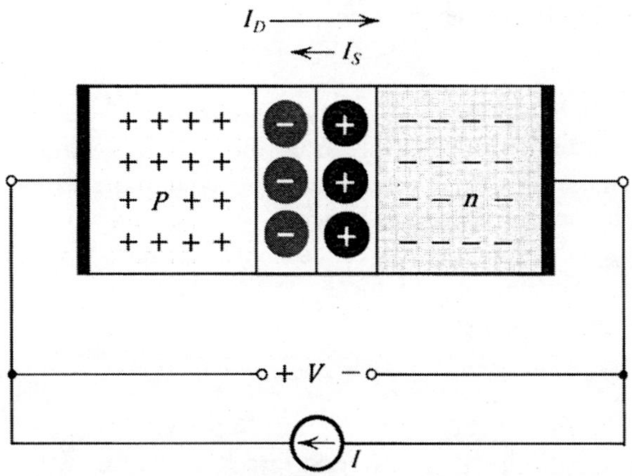

Fig. 1.13 P-N Junction in forward bias

If V_j is the potential drop across the junction, then we can write

$$\frac{n_p}{n_n} = \frac{p_n}{p_p} = exp\left(-\frac{eV_j}{kT}\right)$$

We have that $V_j = V_B - V_{ext}$

Hence

$$\frac{n_p}{n_n} = \frac{p_n}{p_p} = exp\left(-\frac{eV_j}{kT}\right) exp\left(\frac{eV_{ext}}{kT}\right) \quad \ldots\ldots\ldots \quad [1]$$

For zero applied bias-

$$\frac{n_{p_0}}{n_{n_0}} = \frac{p_{n_0}}{p_{p_0}} = exp\left(-\frac{eV_B}{kT}\right) \quad \ldots\ldots\ldots\ldots\ldots [2]$$

By equations [1] and [2], we get-

$$\frac{n_p}{n_n} = \frac{n_{p_0}}{n_{n_0}} exp\left(\frac{eV_{ext}}{kT}\right)$$

$$\frac{p_n}{p_p} = \frac{p_{n_0}}{p_{p_0}} exp\left(\frac{eV_{ext}}{kT}\right)$$

where, n_n is large - plenty of free electrons on ***N-side***

p_p is large - plenty of free holes on P-side

They change marginally with the application f the bias voltage.

i.e. $$n_n \approx n_{n_0}; \quad p_p \approx p_{p_0}$$

$$\frac{n_p}{n_{p_0}} = \frac{p_n}{p_{n_0}} = exp\left(\frac{eV_{ext}}{kT}\right)$$

Thus, $$I = I_0\left[\exp\left(\frac{eV_{ext}}{kT} - 1\right)\right]$$

1.4.2.2. Reverse Bias:

When the polarity of the supply is reversed, the free electrons in the *N-type* are attracted towards the positive terminal, away from the junction, while the electrons from the negative terminal of the supply enter the *P-type* and migrate towards the junction. As a result, the depletion layer becomes wider.

The current flow is minimal and is called reverse current. It is to be noted that minority carriers produce this current, and the device is said to be reverse-biased. Fig. 1.14 shows the reverse biased P-N junction.

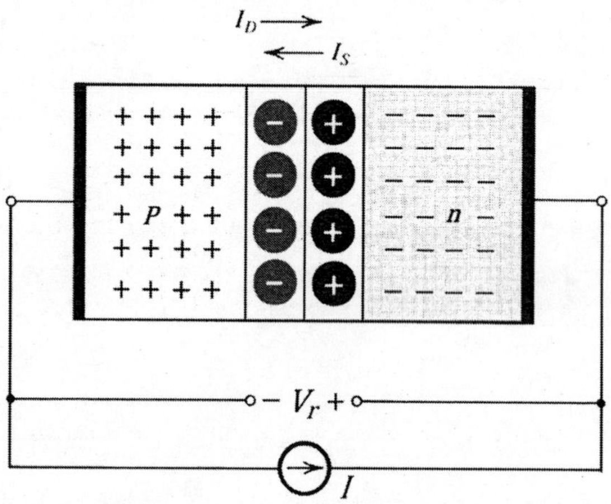

Fig. 1.14. P-N Junction with reverse bias

In reverse bias, the potential drop across the junction is increased. Here current equation is given as-

$$I = I_0\left[1 - \exp\left(-\frac{eV_{ext}}{kT}\right)\right]$$

Thus, in summary, the I-V characteristics of the P-N junction diode are-

$$\Rightarrow \quad I = I_0\left[\exp\left(\frac{eV_{ext}}{kT} - 1\right)\right] \quad \text{Forward bias}$$

$$\Rightarrow \quad I = I_0\left[1 - \exp\left(-\frac{eV_{ext}}{kT}\right)\right] \quad \text{Reverse bias}$$

Since the exponential term is minimal due to negative voltage, it is approximately equal to I_0.

1.4.3. Energy band diagrams for P-N junctions with external bias

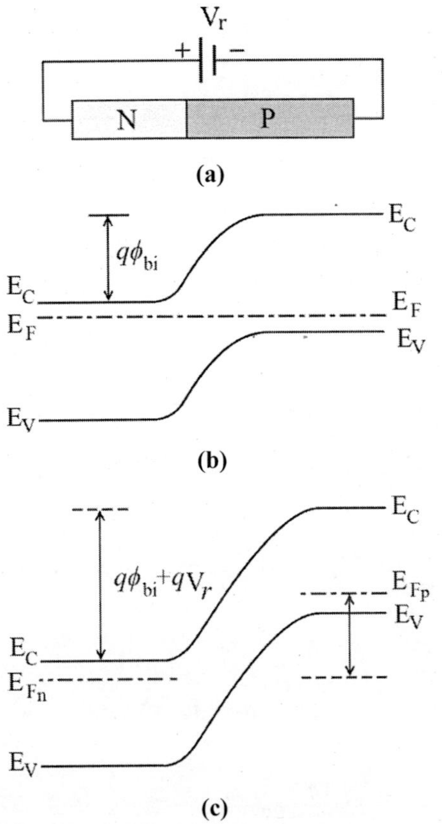

Fig. 1.15: (a) A reverse–biased P-N junction, (b) band diagram without bias, and (c) energy band under reverse bias.

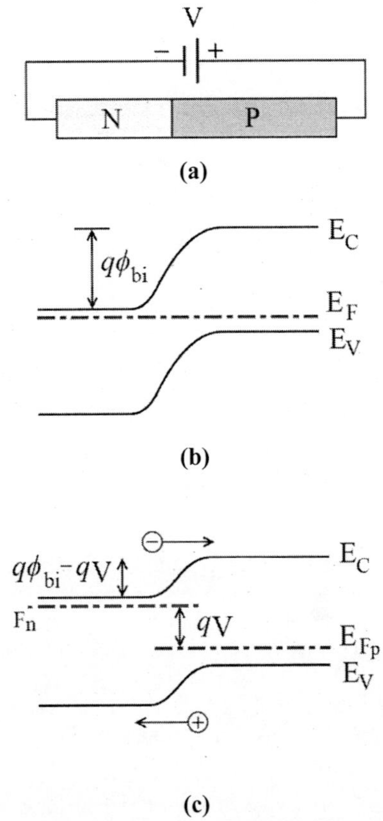

Fig. 1.16: (a) A forward-biased P-N junction, (b) band diagram without bias, and (c) energy band under forward bias.

1.5. Static and dynamic resistance of a P-N junction diode

It is always necessary to remember that no diode can act as an ideal diode. An actual diode does not behave as a perfect conductor when forward–biased and an excellent insulator when reverse-biased. It neither offers zero resistance when forward-biased nor infinite resistance when reverse-biased.

Generally, in a P-N junction diode, there are two types of resistance that are considered as-

1. **Static resistance(R_D):** Static resistance of a P-N junction diode is calculated when the diode is connected to a DC circuit. This resistance is also known as DC resistance or static resistance. It is the ratio of DC voltage across the diode to the DC flowing through it. It is denoted by R_D.

$$R_D = \frac{V_D}{I_D}$$

2. **Dynamic resistance:** Dynamic resistance of a P-N junction diode is defined as the resistance offered by the diode to the AC signal. It is also known as AC resistance. The

dynamic resistance of a diode is equal to the slope of V- I characteristics $\left(\frac{dV}{dI} \ or \ \frac{\Delta V}{\Delta I}\right)$ of the diode. i.e.,

$$\mathrm{r_f} = \frac{\text{Change in voltage}}{\text{Resulted change in current}} = \frac{dV}{dI} = \frac{\Delta V}{\Delta I}$$

Consider the diode current equation

$$I = I_0\left(e^{\frac{V}{\eta V_T}} - 1\right)$$

Differentiating the equation to V, we get dynamic conductance as-

$$g = \frac{dI}{dV} = \frac{I_0 e^{\frac{V}{\eta V_T}}}{\eta V_T}$$

$$\Rightarrow g = \frac{I + I_0}{\eta V_T}$$

$$\Rightarrow r = \frac{\eta V_T}{I + I_0}$$

For a ϕ reverse biased junction $\left(i.e., \frac{V}{\eta V_T} "g\right)$.

g is extremely small, and r is very large. On the other hand, for a forward-biased junction $I > I_0$. So the equation of r becomes

$$r \approx \frac{\eta V_T}{I}$$

From the above expression, it is clear that dynamic resistance is inversely proportional to the forward-biased junction current.

1.6. Diode capacitance

There are usually two types of diode capacitance as-

1. **Depletion layer capacitance or transition capacitance (C_T):**

   The formation of a P-N junction gives rise to the formation of the depletion layer on either side of the junction. Since this capacitance is formed in the junction area, it is also called space charge capacitance. This capacitance arises due to immobile charges at the junction varying with the applied voltage, called the junction capacitance. Fig. 1.17 shows the formation of depletion layer capacitance.

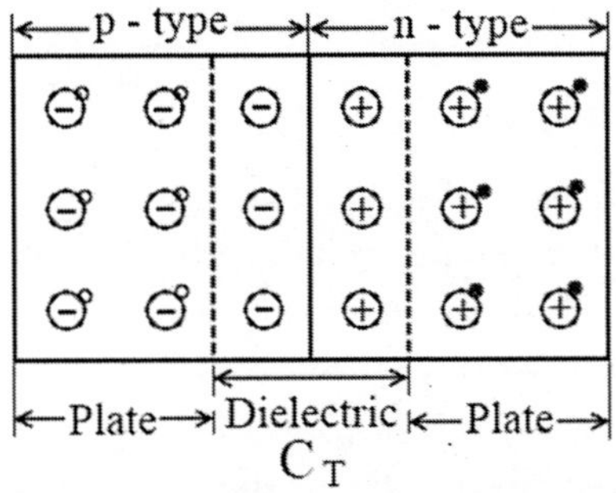

Fig. 1.17: Depletion layer capacitance

The capacitance of a parallel plate capacitor is given as-

$$C = \frac{\epsilon_0 A}{d}$$

Where $\epsilon_0$is the permittivity of dielectric (insulator) between the plate of area A separated by a distance d. Since the depletion layer width (d) increases with the increase in reverse bias voltage, the resulting depletion layer capacitance will decrease with the increased reverse bias.

The depletion layer capacitance depends upon the nature of a P-N junction, semiconductor material, and the magnitude of the applied reverse voltage and is given as -

$$C_T = \frac{C_0}{\left(1 + \frac{V_r}{V_T}\right)^\eta}$$

Where C_0 = Capacitance at zero bias condition

V_r = Applied reverse voltage

V_T = Volt equivalent of temperature

$$\eta = \begin{cases} \frac{1}{2} & \text{for a step or abrupt alloy junction} \\ \frac{1}{3} & \text{for linear graded junction or diffused junction} \end{cases}$$

It is evident from the above equations that the value of depletion layer capacitance (C_T) can be controlled by varying the applied reverse voltage. This property of variable capacitance, possessed by a reverse biased P-N junction, is used in constructing a device called varicap or varactor for FM circuits. The range of depletion layer capacitance is approximately in nF(nanofarad).

2. **Diffusion or storage capacitance (C_D)**

The capacitance in a forward-biased junction is called diffusion or storage capacitance. It is different from the depletion layer capacitance in a reverse-biased junction. The diffusion capacitance arises due to the arrangement of minority carrier density. The value of diffusion capacitance is much larger than the depletion layer capacitance ($i.e.,$ C_D"C_T). The following expression provides its value for abrupt junction:

$$C_D = \frac{\tau \,.\, I}{\eta \,.\, V_T}$$

Where τ = Lifetime of the carriers

I = Forward current

V_T = Volt equivalent of temperature

$$\eta = \text{constant} = \begin{cases} 1 & \text{for } Ge \\ 2 & \text{for } Si \end{cases}$$

It is clear from the equation that the diffusion capacitance is directly proportional to the forward current (I). Consider a forward-biased P-N junction carrying a current of (I) amperes through it. Suppose the junction is suddenly reverse-biased. It caused the forward current to reduce quickly to zero. But it leaves several majority charge carriers stored within the depletion region. This charge represents a stored charge in the reverse biased condition and must be removed from the space charge region.

The removal of stored charge takes a finite time. It represents an effect similar to the discharging of a capacitor. Thus the quantity of the stored charge represents the magnitude of diffusion capacitance. It noted that the impact of diffusion capacitance is negligible for a reverse-biased P-N junction.

1.7. Temperature Effect on Diode

Fig. 1.18 shows the effect of temperature on diode characteristics.

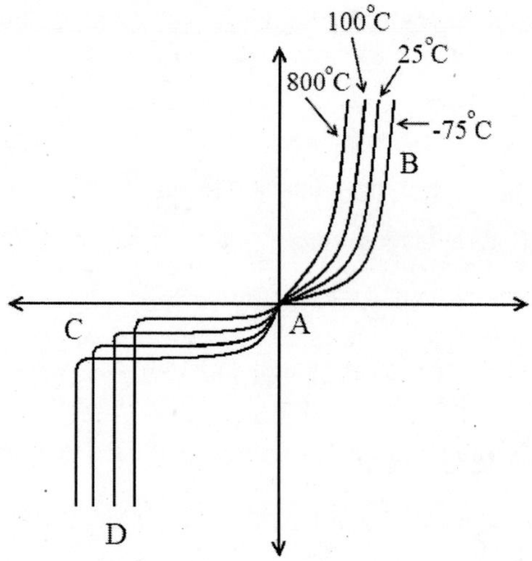

Fig. 1.18: Diode characteristics for different temperatures

A-B curve: This curve shows diode characteristics for different temperatures in the forward bias. It is evidenced from Fig.1.18 that the curve moves towards the left with the increase in the temperature. It indicates that the conductivity of semiconductors increases with an increase in temperature.

The intrinsic concentration (n_i) of the semiconductors is dependent on temperature as given by:

$$n_i^2 = KT^3 e^{\frac{-Eg}{KT}}$$

where Eg = the energy gap

K = a voltage man constant

T = a constant independent of temperature

When the temperature is high, the electrons of the outermost shell take the thermal energy and become free. So conductivity increases with temperature. Hence with an increase in temperature, the A-B curve would shift towards the left, i.e., the curve would rise sharply, and the breakdown voltage would also decrease with an increase in temperature.

A-C curve: This curve shows diode characteristics in the reverse biased region until the breakdown voltage for different temperatures. We know n_i concentration would increase with an increase in temperature, and hence minority charges would increase with an increase in temperature.

The minority charge carriers are also known as thermally generated carriers, and the reverse current depends on minority carriers only. Hence as the number of minority charge carriers increases, the reverse current would also increase with temperature, as shown in the figure.

The reverse saturation current gets doubles with every $10°C$ increase in temperature.

C-D curve: This curve shows the characteristics of a diode in a reverse biased region from the breakdown voltage point onwards. As with an increase in temperature, loosely bonded electrons are already free, and freeing the other electrons would take more voltage than earlier. Hence breakdown voltage increases with an increase in temperature, as depicted in Fig. 1.18.

1.8. Switching time of a Diode

The switching time of a diode is defined as the time it takes to change its state from a forward-biased to a reverse-biased state. In other words, the forward current through the diode doesn't reduce to reverse saturation current as the reverse voltage is applied. Instead, it takes time for the current to reduce from forward to reverse saturation current. This time is also known as reverse recovery time.

To discuss the switching time, let's analyze what would happen when we change the diode state from forward bias to reverse bias. This state change takes time which is known as reverse recovery time. First, consider the diode circuit to analyze the diode's switching time, as shown in Fig. 1.19.

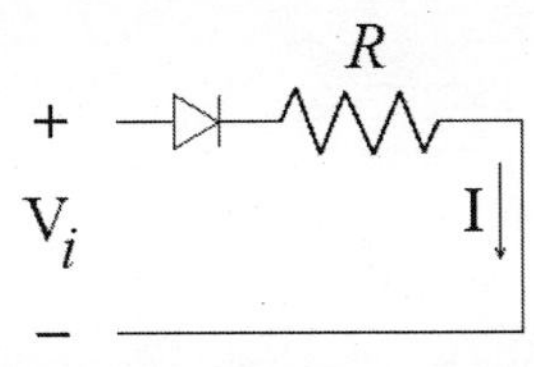

Fig. 1.19: A diode circuit

Initially, when the voltage applied is +V, the diode is in a forward bias state. Hence, there are many minority carriers near the junction. Then, there is an exponential decrease in the concentration of minority carriers and a continuous flow of majority carriers across the junction. Hence, let us assume the current as ***I*** in the forward bias.

Later, the applied voltage is changed to –V at time $t=t_1$, i.e., the diode is reverse biased. With the instant change in the operation state to reverse bias, the minority carriers start moving in the opposite direction. As a result, the current flow remains the same with a change in direction. However, the high reverse current continues for a shorter duration. As a result, the concentration of the stored minority carriers starts decreasing, and the current also starts decreasing exponentially, as shown in Fig. 1.20.

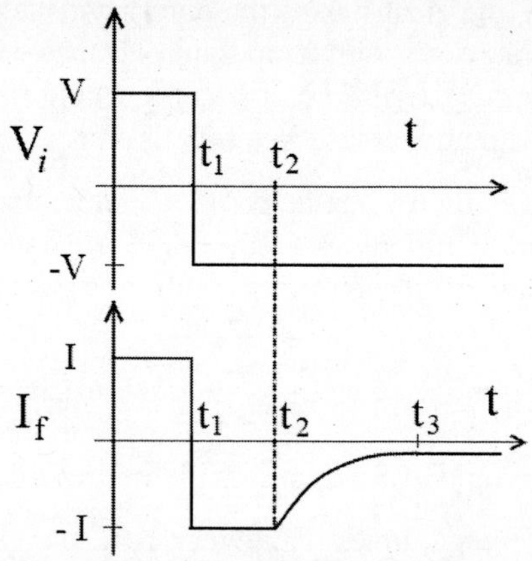

Fig. 1.20: Switching timing of diode

The time gap $t_2 - t_1$ in which the reverse current is high (i.e., equal to I) is known as storage time, and the time gap from t_2 to t_3, i.e., the time reverse current becomes equivalent to reverse saturation current, is known as transient time. The total time from t_1 to t_3 is known as reverse recovery time.

Storage time is the time required for majority charge carriers to return to their initial position when the diode is suddenly reverse biased from the forward biased condition". Fig. 1.21 illustrates the concept of the storage time of the diode.

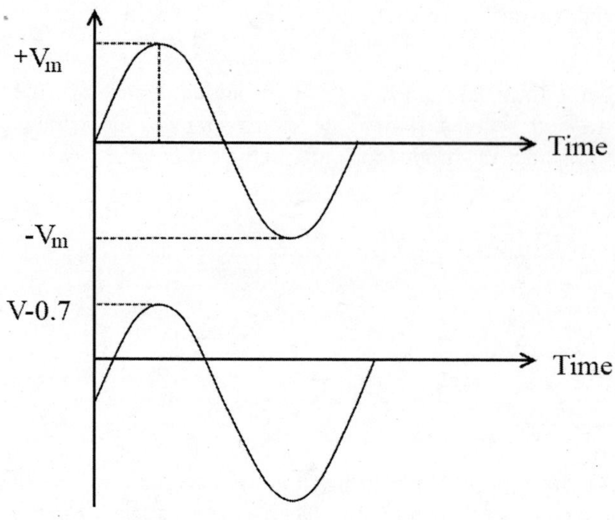

Fig. 1.21: Storage timing of diode

1.9. DC Power Supply

The DC power supply is a multipurpose circuit used in all electronic devices. The basic principle of a DC power supply is to convert the input AC supply into a DC output voltage. Fig. 1.22 shows the block diagram of a DC power supply.

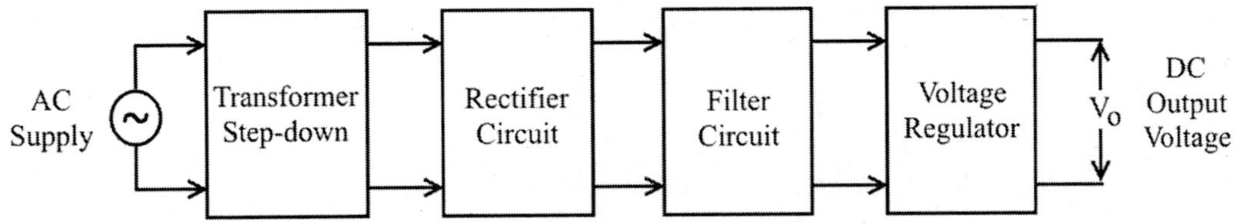

Fig. 1.22: Block diagram of DC power supply

It consists of a power transformer, rectifiers, filters, and voltage regulators. The power transformer steps down the application's input to the desired voltage. The Rectifier converts the ac voltage to a pulsating DC voltage. Finally, the regulator maintains a constant output voltage.

2. Diode as a Rectifier

The process of converting an AC signal into a DC signal is called rectification. A rectifier is an electronic device that offers low resistance to current in one direction and high resistance to current in the opposite direction. There are two types of rectifiers: (a) Half Wave rectifiers and (b) Full wave rectifiers.

2.1. Half wave Rectifier

A halfwave rectifier is a circuit in which the positive or negative half of the AC wave is allowed to pass while the other half is blocked. As a result, the output voltage is low as only half of the input waveform reaches the output. Fig. 1.23 shows a halfwave rectifier circuit and its output waveform.

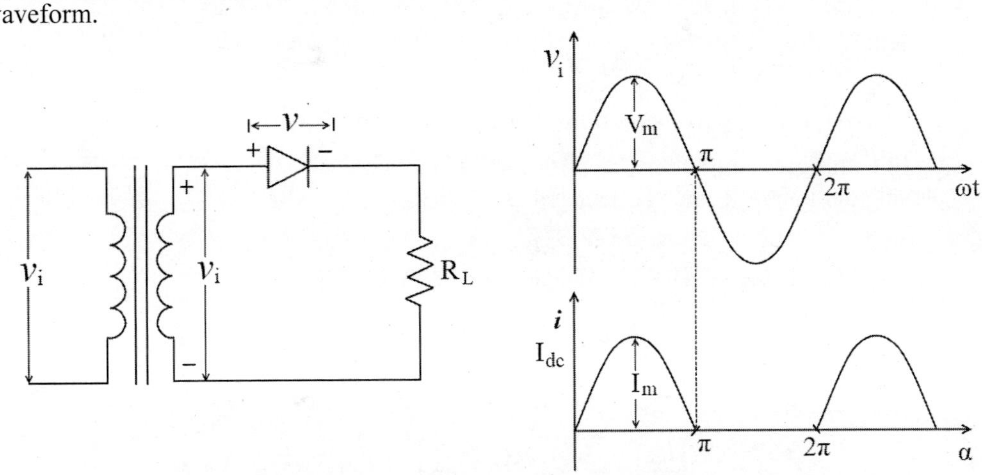

Fig.1.23: Half Wave Rectifier circuit and its output waveform

Operation: Equivalent circuit of half wave rectifier for analysis

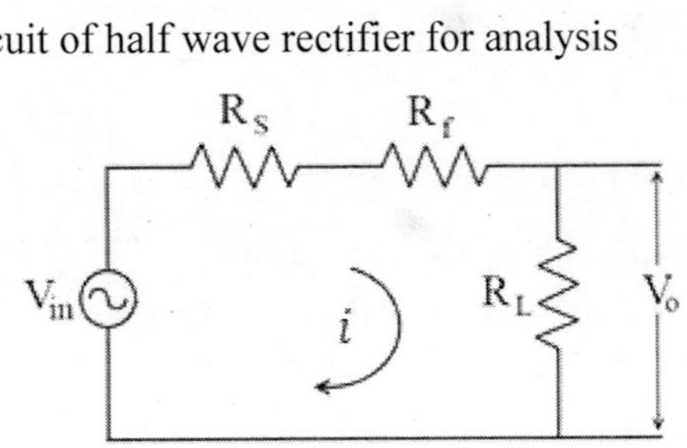

In the positive half cycle of thc input voltage, the diode conducts and current i flows through R_L.

$$i = I_m \sin\alpha \; ; \quad 0 \le \alpha \le \pi \ldots\ldots\ldots\ldots\ldots (1)$$

During the negative half cycle, the diode is reverse biased, and no current flows through the diode or load R_L

$$\therefore\ i = 0 \; ; \quad \pi \le \alpha \le 2\pi \ldots\ldots\ldots\ldots\ldots\ldots (2)$$

where $\alpha = \omega t$

$$\therefore \text{ Peak Value load Current } I_m = \frac{V_m}{R_f + R_L}$$

Where R_f = dynamic resistance of the diode.

R_L = Load resistance.

(a) DC output current (I_{dc}): A DC ammeter indicates the average value of the current passing through it. The average value of a function is given by the area of one cycle of the curve divided by the base.

$$\therefore I_{dc} = \frac{1}{2\pi}\int_0^{\pi} i \cdot d\alpha \ldots\ldots\ldots\ldots (3)$$

For Half Wave Rectifier

$$I_{dc} = \frac{1}{2\pi}\int_0^{\pi} I_m \sin\alpha \, d\alpha = \frac{I_m}{2\pi}[-\cos\alpha]_0^{\pi} \ldots\ldots\ldots (4)$$

$$I_{dc} = \frac{I_m}{\pi}$$

Limit 0 to π is considered as current flows in this period only.

(b) RMS Current ac ammeter (I_{rms}): An AC ammeter indicates the effective or rms current passing through it. It can be calculated as-

$$\therefore\ I_{rms} = \left[\frac{1}{2\pi}\int_0^{2\pi} i^2 d\alpha\right]^{\frac{1}{2}} \ldots\ldots\ldots\ldots\ldots\ldots (5)$$

$$= \left[\frac{I_m^2}{2\pi}\int_0^{\pi} \frac{1-\cos 2\alpha}{2} d\alpha\right]^{\frac{1}{2}}$$

$$= \left[\frac{I_m^2}{4\pi}\left[\alpha - \frac{\sin 2\alpha}{2}\right]_0^{\pi}\right]^{\frac{1}{2}} = \left[\frac{I_m^2}{4\pi}(\pi)\right]^{\frac{1}{2}}$$

$$I_{rms} = \frac{I_m}{2} \ldots\ldots\ldots\ldots\ldots\ldots\ldots (6)$$

Note: R.M.S. value of sinusoidal wave is $\frac{I_m}{\sqrt{2}}$

(c) DC Voltage: It indicates the average value of the voltage across its terminals. If the voltmeter is connected across a diode and the diode is conducting, it has resistance R_f and the voltage across it is iR_f. For a non-conducting diode, the current is zero, and voltage V_i appears across the diode.

$$\therefore v = iR_f = I_m R_f \sin\alpha \qquad 0 \le \alpha \le \pi$$

$$v = V_m \sin\alpha \, ; \quad \pi \le \alpha \le 2\pi \ldots\ldots\ldots\ldots\ldots\ldots\ldots\ldots (7)$$

$\therefore$ DC Voltmeter reading is given as-

$$V_{dc} = \frac{1}{2\pi}\left(\int_0^{\pi} I_m R_f \sin\alpha\, d\alpha + \int_0^{\pi} V_m \sin\alpha\, d\alpha\right)$$

$$= \frac{1}{\pi}\left[I_m R_f - I_m(R_f + R_L)\right] \qquad \left[\because I_m = \frac{V_m}{R_f + R_L}\right]$$

$$V_{dc} = -\frac{I_m R_L}{\pi} \ldots\ldots\ldots\ldots (8)$$

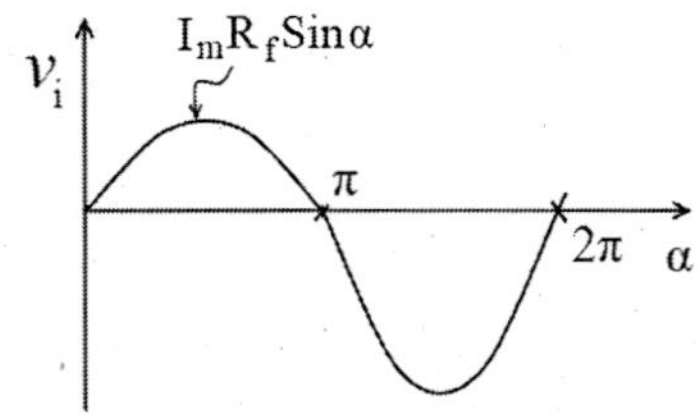

As $\quad I_{dc} = \dfrac{I_m}{\pi}$

$\therefore$ DC diode voltage $= -I_{dc}R_L$

(d) Reading of a wattmeter: It indicates the average value of the product of the instantaneous current through its current coil and the instantaneous voltage across its potential coil.

$$\therefore\ P_i = P_{ac} = \frac{1}{2\pi}\int_0^{\pi} V_i\, i\ d\alpha \ldots\ldots\ldots\ldots\ldots\ldots (9)$$

Now $\quad V_i = i(R_f + R_L) \quad$ for $0 \le \alpha \le \pi$

$$P_{dc} = I_{dc}^2 \times R_L = \left(\frac{I_m}{\pi}\right)^2 R_L$$

$$\therefore\ P_i = \frac{1}{2\pi}\int_0^{\pi} i^2 (R_f + R_L) d\alpha$$

$$= \frac{1}{2\pi}\int_0^{\pi} I_m^2 \sin^2\alpha\,(R_f + R_L) d\alpha$$

$$= \frac{I_m^2}{4}(R_f + R_L) \qquad \text{now as } \left(\frac{1}{2\pi}\int_0^{\pi} I_m^2 \sin^2\alpha\, d\alpha\right)^{\frac{1}{2}} = \frac{I_m}{2}$$

$$P_{ac} = P_i = I_{rms}^2(R_f + R_L) \ldots\ldots\ldots\ldots\ldots\ldots\ldots (10)$$

(e) Peak Inverse Voltage (PIV): It refers to the maximum voltage a diode can withstand in the reverse-biased direction before breakdown.

PIV for H.W.R. $= V_m$

(f) Regulation: It is the percentage change in the output voltage from no-load to full-load.

$$\%\ \text{regulation} = \frac{V_{\text{No Load}} - V_{\text{Full Load}}}{V_{\text{No Load}}} \ldots\ldots\ldots\ldots\ldots (11)$$

The variation of V_{dc} and I_{dc} for the Half wave rectifier is-

$$I_{dc} = \frac{I_m}{\pi} = \frac{V_m/\pi}{R_f + R_L} \ldots\ldots\ldots\ldots\ldots\ldots\ldots (12)$$

$$V_{dc} = I_{dc}R_L = \frac{V_m}{\pi} - I_{dc}R_f \ldots\ldots\ldots\ldots (13)$$

$$\left[V_{NL} = \frac{V_m}{\pi}\right]$$

$$I_{dc}R_L + I_{dc}(R_f + R_L) = \frac{V_m}{\pi}$$

$$\therefore V_{dc} = \frac{V_m}{\pi} - I_{dc}R_f = V_{FL}$$

The Thevenin's equivalent circuit for Half Wave Rectifier is as shown below.

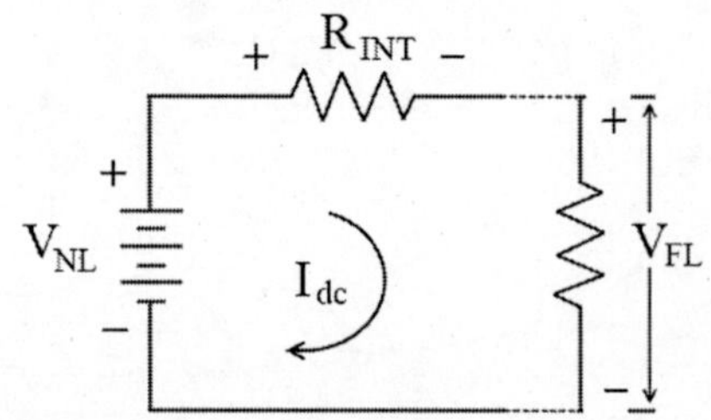

Fig. 1.24(a): Thevenin's equivalent of Half wave rectifier

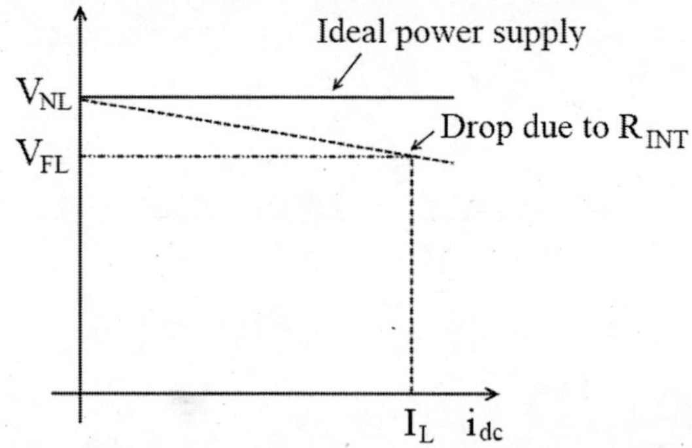

Fig. 1.24(b): Variation in terminal voltage with load current for ideal & practical power supply

(g) Ripple Factor (*r*): A rectifier should convert AC into DC. However, the output of a halfwave rectifier is not constant (pure DC). Instead, it has periodically fluctuating components remaining in the output wave. The measure of the fluctuating components is given by the ripple factor 'r.'

The ripple factor is defined as the ratio of the rms value of the alternating signal component to the average signal value.

$$\text{r} = \frac{\text{the rms value of alternating components of a wave}}{\text{the average value of a wave}}$$

$$\frac{I_{r,rms}}{I_{dc}} = \frac{V_{r,rms}}{V_{dc}} \quad \ldots\ldots\ldots\ldots\ldots\ldots..(14)$$

Where $I_{r,rms}$ and $V_{r,rms}$ denotes the *rms* value of DC components of the current and voltage, respectively.

The instantaneous ac components of current

$$i = i - I_{dc}$$

$$\therefore \ I_{r,rms} = \sqrt{\frac{1}{2\pi}\int_0^{2\pi} (i - I_{dc})^2 \, d\alpha}$$

$$= \sqrt{\frac{1}{2\pi}\int_0^{2\pi}(i^2 - 2iI_{dc} + I_{dc}^2)\, d\alpha}$$

The first term of integral is simply I_{rms}^2 of the total wave. The second term is

$$(-2I_{dc})I_{dc} = -2I_{dc}^2 \qquad \left[\because \frac{1}{2\pi}\int_0^{2\pi} i\, d\alpha = I_{dc}\right]$$

∴ R.M.S. ripple current

$$I_{r,rms} = \sqrt{I_{rms}^2 - 2I_{dc}^2 + I_{dc}^2} = \sqrt{I_{rms}^2 - I_{dc}^2}$$

Hence,

$$r = \frac{\sqrt{I_{rms}^2 - I_{dc}^2}}{I_{dc}} = \sqrt{\left(\frac{I_{rms}}{I_{dc}}\right)^2 - 1}$$

For half wave rectifier

$$\frac{I_{rms}}{I_{dc}} = \frac{I_m/2}{I_m/\pi} = 1.57$$

$$\therefore\ r = \sqrt{(1.57)^2 - 1} = 1.21$$

$$r = 121\%$$

⇒ *rms* ripple voltage exceeds DC output voltage.

∴ A Halfwave rectifier is a poor circuit for rectification.

(h) The ratio of Rectification or Efficiency (η)

$$\eta = \frac{\text{DC power output}}{\text{ac input power}} = \frac{P_{o(dc)}}{P_{i(ac)}}$$

$$P_{o(dc)} = I_{dc}^2 R_L = \left(\frac{I_m}{\pi}\right)^2 R_L$$

$$P_{i(ac)} = I_{rms}^2(R_f + R_L) = \left(\frac{I_m}{2}\right)^2 (R_f + R_L)$$

$$\therefore\ \eta = \frac{\frac{I_m^2}{\pi^2} R_L}{\frac{I_m^2}{4}(R_f + R_L)} = \frac{4}{\pi^2}\cdot\frac{1}{1 + \frac{R_f}{R_L}} \text{ for max } \eta,\ \ R_f \parallel R_L$$

⇒ Only 40.6% of ac power input is converted into DC power input.

(i) Transformer utilization Factor (TUF): The ratio of output DC power delivered to the load to the ac rating of transformer secondary.

$$TUF = \frac{P_{0(dc)}}{a.c.\text{ rating of the transformer secondary}}$$

Now, AC rating of transformer secondary $= V_{rms}\cdot I_{rms} = \frac{V_m}{\sqrt{2}} \times \frac{I_m}{2}$

$$= \frac{I_m^2}{2\sqrt{2}}(R_f + R_L + R_S)$$

Where R_S = resistance of transformer secondary winding

$$\therefore\ TUF = \frac{\frac{I_m^2}{\pi^2}\cdot R_L}{\frac{I_m^2(R_S + R_f + R_L)}{2\sqrt{2}}}$$

$$TUF = \frac{2\sqrt{2}}{\pi^2}\cdot\frac{1}{1+\frac{R_s + R_f}{R_L}}$$

T.U.F. will be maximum when R_L is very large compared to $R_s + R_f$ and is given by,

$$TUF_{max} = 0.286$$

Disadvantages of Half wave rectifier

(i) Excessive ripple (1.21)

(ii) Very low efficiency (0.406)

(iii) Less T.U.F. (0.286)

(iv) Rectifies only half of the input

2.2. Full Wave Centre tapped rectifier

A full-wave rectifier converts the input waveform to one of its output's constant polarity (positive or negative). It converts both polarities of the input waveform to pulsating DC and yields a higher average output voltage. Fig.1.25 shows the centertap full wave rectifier circuit.

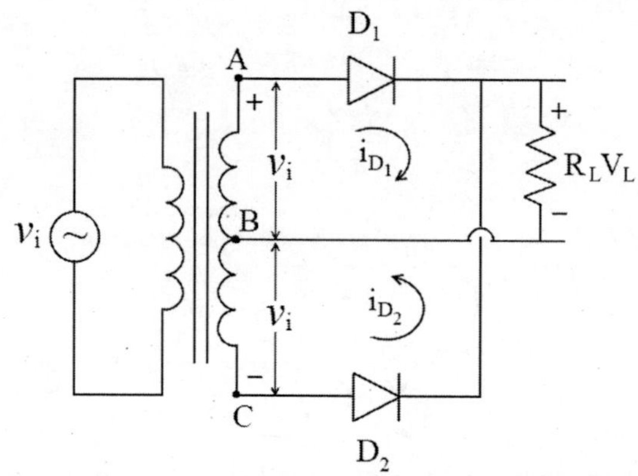

Fig. 1.25: A Centertap Full Wave Rectifier Circuit

A full wave rectifier comprises two halfwave rectifiers. During the positive half cycle of input, the upper transformer secondary winding is positive with respect to the center tap. Hence diode D_1 conducts. The lower transformer secondary winding is negative with respect to the center tap; hence, D2 is off(open).

During a negative half cycle, reverse action takes place & diode D_2 conducts, and D_1 is off (open). Hence upper half of the transformer's secondary winding is disconnected from the load.

$$\therefore \left.\begin{array}{l} id_1 = I_m \sin \omega t \\ id_2 = 0 \end{array}\right\} ; \qquad 0 \le \omega t \le \pi \ldots\ldots\ldots\ldots (1)$$

and

$$\therefore \left.\begin{array}{l} id_1 = 0 \\ id_2 = I_m \sin \omega t \end{array}\right\} ; \qquad \pi \le \omega t \le 2\pi \ldots\ldots\ldots\ldots (2)$$

(a) Average value:

$$I_{av} = I_{dc} = \frac{1}{T}\int_0^T i\, dt$$

$$= \frac{1}{\pi}\int_0^{\pi} I_m \sin \alpha\, d\alpha = \frac{I_m}{\pi}[-\cos \alpha]_0^{\pi}$$

$$I_{dc} = \frac{2I_m}{\pi}$$

$$I_{dc} = I_{av} = \frac{2V_m}{\pi(R_f + R_L)} \ldots\ldots\ldots\ldots\ldots (3)$$

(b) R.M.S. Value:

$$I_{rms} = \left[\frac{1}{T}\int_0^T i^2\, dt\right]^{\frac{1}{2}}$$

$$= \left[\frac{1}{\pi}\int_0^{\pi} I_m^2 \sin^2 \alpha\, d\alpha\right]^{\frac{1}{2}} = \left[\frac{I_m^2}{\pi}\int_0^{\pi} \frac{1 - \cos 2\alpha}{2} d\alpha\right]^{\frac{1}{2}}$$

$$= \left[\frac{I_m^2}{2\pi}\left[\alpha - \frac{\sin 2\alpha}{2}\right]_0^{\pi}\right]^{\frac{1}{2}}$$

$$I_{rms} = \frac{I_m}{\sqrt{2}} \ldots\ldots\ldots\ldots\ldots\ldots\ldots\ldots (4)$$

(c) Output DC voltage:

$$I_m = \frac{V_m}{R_s + R_f + R_L}$$

Where R_s = resistance of a secondary winding of the transformer

R_f = Forward resistance of a diode

R_L = Load Resistance

$$\therefore V_{dc} = I_{dc} \cdot R_L = \frac{2I_m}{\pi} \cdot R_L$$

$$\therefore V_{dc} = \frac{\frac{2V_m}{\pi}}{1 + \frac{R_s + R_f}{R_L}} \ldots\ldots\ldots\ldots\ldots (6)$$

Thus practical DC voltage is less than the theoretical value due to a finite drop across R_s and R_f. It can be approximated to theoretical value by keeping R_L very large as compared to $R_s + R_f$. It can be given as-

$$I_{dc} = \frac{2I_m}{\pi} = \frac{\frac{2V_m}{\pi}}{R_s + R_f + R_L}$$

$$\therefore I_{dc}(R_s + R_f) + I_{dc}R_L = \frac{2V_m}{\pi}$$

$$\therefore V_{dc} = \frac{2V_m}{\pi} - I_{dc}(R_s + R_f)$$

(d) Efficiency: (i) $P_{i(ac)} = I_{rms}^2(R_s + R_f + R_L)$

$$= I_m^2(R_s + R_f + R_L)$$

(ii) $P_{o(dc)} = I_{dc}^2 \cdot R_L$

$$= \frac{4I_m^2}{\pi^2} \cdot R_L$$

$$\therefore \quad \% \eta = \frac{P_{o(dc)}}{P_{i(ac)}} \times 100 = \frac{4I_m^2 \cdot R_L}{\frac{I_m^2}{2}(R_s + R_f + R_L)} \times 100$$

$$\Rightarrow \% \eta = \frac{8}{\pi^2\left(1 + \frac{R_s + R_f}{R_L}\right)} \times 100$$

Maximum efficiency is achieved when R_L is very large compared to $(R_s + R_f)$

$$\therefore \% \eta_{\max} = \frac{8}{\pi^2} = 81.2\%$$

(e) Ripple factor:

$$r = \sqrt{\left(\frac{I_{rms}}{I_{ds}}\right)^2 - 1} = \sqrt{\frac{\frac{I_m^2}{2}}{\frac{4I_m^2}{\pi^2}} - 1}$$

$$r = \sqrt{\frac{\pi^2}{8} - 1} = 0.48$$

(f) Transformer utilization factor (T.U.F.):

AC rating secondary

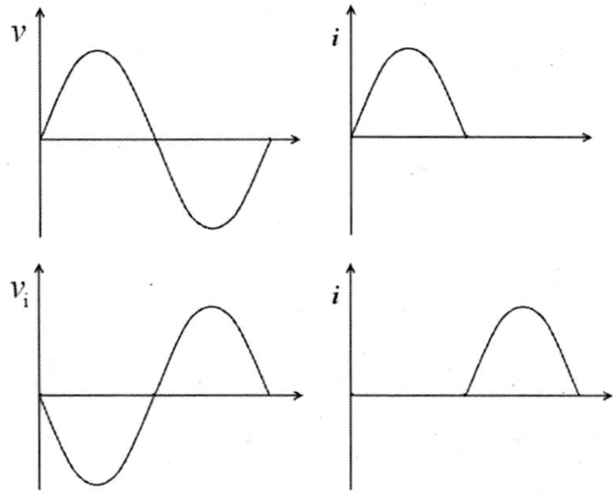

$$= \frac{V_m}{\sqrt{2}} \cdot \frac{I_m}{2} + \frac{V_m}{\sqrt{2}} \cdot \frac{I_m}{2}$$

$$= \frac{2V_m I_m}{2\sqrt{2}}$$

$$= \frac{V_m I_m}{\sqrt{2}}$$

$$= \frac{I_m^2}{\sqrt{2}}(R_s + R_f + R_L)$$

$$\therefore \quad TUF = \frac{P_{o(dc)}}{a.c. \text{ rating of Transformer secondary}}$$

$$= \frac{\frac{4I_m^2}{\pi^2} \cdot R_L}{\frac{I_m^2}{\sqrt{2}}(R_s + R_f + R_L)}$$

$$\therefore \quad TUF = \frac{4\sqrt{2}}{\pi^2\left(1 + \frac{R_s + R_f}{R_L}\right)}$$

T.U.F. is maximum where R_L is very large compared to $R_s + R_f$

$$\therefore \; TUF_{(max)} = 0.579$$

(g) PIV: When V_I is applied, during the negative half cycle, D_1 is OFF, and D_2 conducts. The peak voltage that occurs across R_L is V_m. At the same time peak voltage from B to A is also V_m.

∴ PIV = sum of the two is the voltage across D_1

∴ PIV = $2V_m$.

Advantages of a centertap full wave rectifier

(i) Efficiency is high → 81.2%

(ii) Ripple factor is low → 0.48

(iii) TUF increases to 57.9%

(iv) It rectifies the full cycle.

Disadvantages:

(i) It requires a centre tap transformer.

(ii) PIV is $2V_m$

2.3. Bridge Rectifier

Bridge Rectifier is a full-wave rectifier circuit that can be fed directly from the line. Therefore, it does not require a high PIV rating of diodes compared to a centre-tapped full-wave rectifier. Fig. 1.26 shows the circuit diagram for a bridge rectifier.

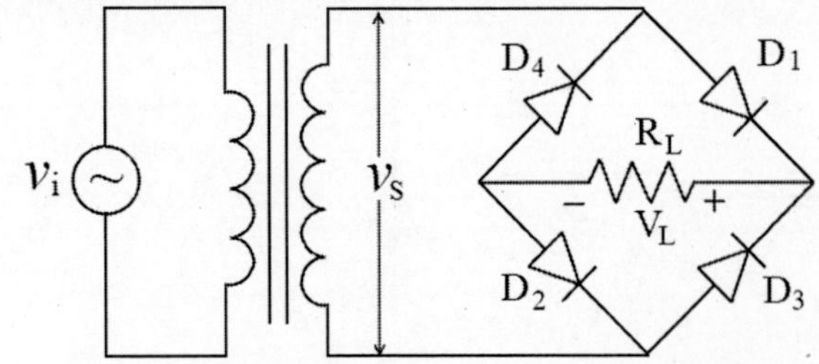

Fig. 1.26: Full Wave Bridge Rectifier circuit

During the positive half cycle of V_s, diode D_1 and D_2 conduct, mounting an output voltage across R_L. Diodes D_3 and D_4 are reverse biased.

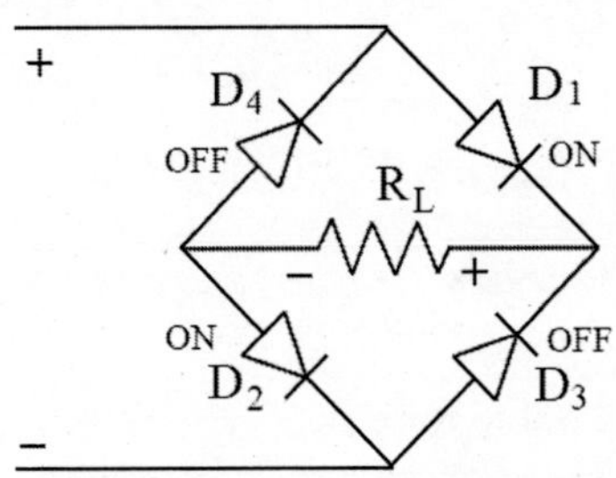

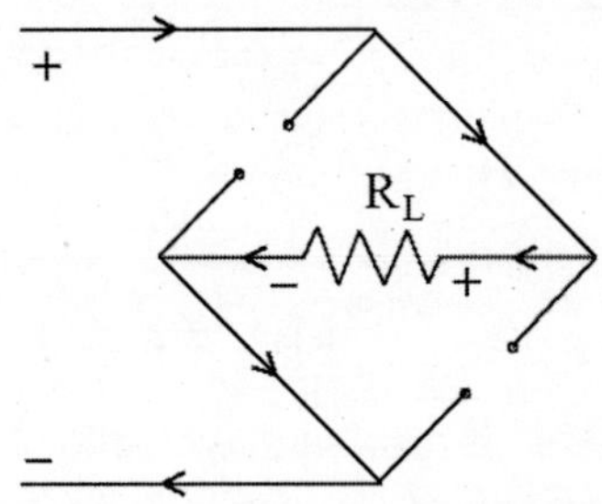

During the negative half cycle D_1 and D_2 are open, and D_3 and D_4 conducts, i.e., forward biased.

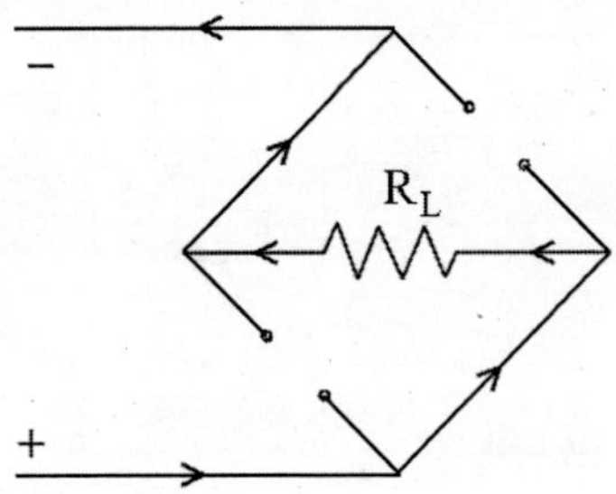

(a) DC Output Current

$$I_{dc} = \frac{2I_m}{\pi} \quad \ldots\ldots\ldots\ldots\ldots\ldots\ldots (1)$$

Where $$I_m = \frac{V_m}{2R_f + R_L + R_s} \quad \ldots\ldots\ldots\ldots (2)$$

$2R_f$ is used since two diodes are conducted at the same time. Ripple factor and efficiency are the same as center-tapped fullwave rectifiers.

(b) PIV: Peak voltage across diode = V_m

(c) $I_{rms} = \frac{I_m}{\sqrt{2}}$

(d) Output DC voltage V_{dc}

$$V_{dc} = \frac{\frac{2V_m}{\pi}}{1 + \frac{R_s + 2R_f}{R_L}}$$

$$V_{dc} = \frac{2V_m}{\pi} - I_{dc}(2R_f + R_s)$$

(e) Efficiency, η

$$\eta = \frac{0.812}{1 + \frac{R_s + 2R_f}{R_L}}$$

$$\%\ \eta = 81.2$$

(f) Ripple factor, $r = 0.48$

(g) T.U.F. AC rating of the transformer secondary-

$$\frac{V_m}{\sqrt{2}}\frac{I_m}{\sqrt{2}} = \frac{I_m^2(R_s + 2R_f + R_L)}{2}$$

$$P_{o(dc)} = \frac{4I_m^2}{\pi^2} \cdot R_L$$

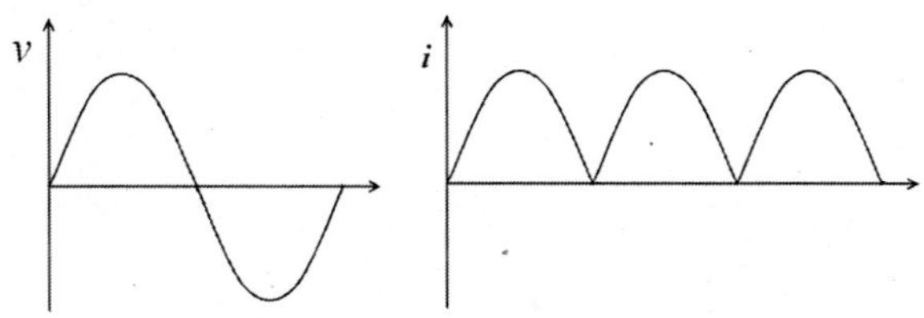

$$\therefore\ TUF = \frac{\frac{4I_m^2}{\pi^2} \cdot R_L}{\frac{I_m^2}{2}\left(R_s + R_{2f} + R_L\right)}$$

$$TUF = \frac{8}{\pi^2\left(1 + \frac{R_s + 2R_f}{R_L}\right)}$$

$$TUF\ (max) = 0.812 \qquad (\because R_s \| R_L) \because R_s + 2R_f \| R_L$$

Advantages

(i) No center tap transformer is required

(ii) Efficiency is high → 81.2%

(iii) Ripple factor is low → 0.48

(iv) PIV is only V_m

(v) TUF increases to 81.2%

Disadvantages:

(i) It requires four diodes

2.4. Filters

The purpose of a rectifier is to convert AC into DC. However, the output of the half wave and full wave rectifier contains ripple components in addition to the DC, i.e., the output is a pulsating DC. Hence filters are included between the Rectifier and load to attenuate these ripple components.

2.4.1 Inductor filter

Fig. 1.26(a) shows an inductor filter. It consists of a choke in series with the load. The characteristic of an inductor is to oppose any change in current. Hence it has a smooth output, as shown in Fig. 1.26(b).

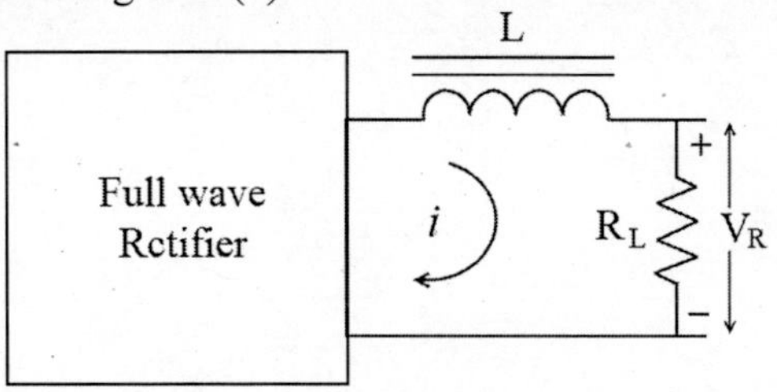

Fig. 1.26(a): Inductor filter circuit

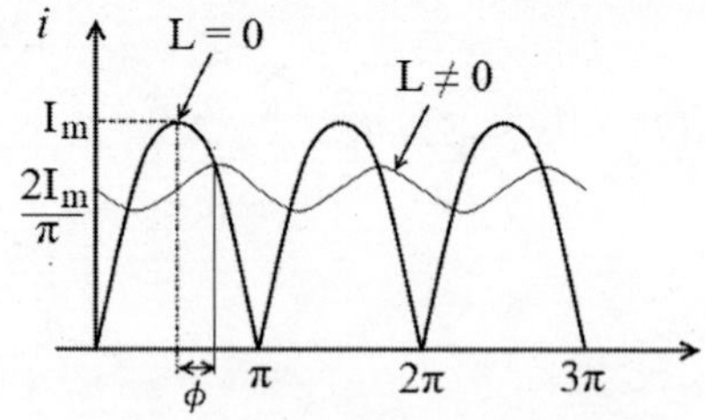

Fig. 1.26(b): Output of inductor filter

For an input voltage $V_s = V_m \sin \omega t$

The output current is given as-

$$I = \frac{2I_m}{\pi} - \frac{4I_m}{3\pi}\cos(2\omega_o t - \phi) + \frac{7I_m}{15\pi}\cos(4\omega_o + \phi)$$

Assuming that the third and fourth harmonic contribute little output, the output current is

$$I = \frac{2I_m}{\pi} - \frac{4I_m}{3\pi}\cos(2\omega_o t - \phi) \ldots\ldots\ldots\ldots\ldots (1)$$

Neglecting diode, choke and transfer resistance compared with R_L for simplification.

$$\therefore \text{ The DC component of current } I_m = \frac{V_m}{R_L}$$

Where $V_m =$ no load peak voltage of V_s,

Now the impedance of L and R_L in series with the AC component has a magnitude

$$Z = \sqrt{R_L^2 + (2\omega L)^2}$$

$$Z = \sqrt{R_L^2 + 4\omega^2 L^2} \ldots\ldots\ldots\ldots\ldots\ldots (2)$$

$\therefore$ For the AC component,

$$I_m = \frac{V_m}{\sqrt{R_L^2 + 4\omega^2 L^2}} \ldots\ldots\ldots\ldots\ldots\ldots (3)$$

The load current becomes,

$$I = \frac{2V_m}{\pi R_L} - \frac{4V_m}{3\pi\sqrt{R_L^2 + 4\omega^2 L^2}}\cos(2\omega t - \phi) \ldots\ldots\ldots\ldots (4)$$

Where ϕ is the angle by which load current lags behind the voltage

$$\phi = \tan^{-1}\frac{2\omega L}{R_L} \ldots\ldots\ldots\ldots\ldots (5)$$

Ripple Factor: $r = \frac{\text{the rms value of the AC component}}{\text{DC value of the wave}} = \frac{I_r, rms}{I_{dc}}$

$$= \frac{\frac{4V_m}{3\pi\sqrt{2}\sqrt{R_L^2 + 4\omega^2 L^2}}}{\frac{2V_m}{\pi R_L}} \quad \ldots\ldots\ldots\ldots\ldots\ldots (7)$$

$$= \frac{2}{3\sqrt{2}} \times \frac{1}{\sqrt{1 + \frac{4\omega^2 L^2}{R_L^2}}} \quad \ldots\ldots\ldots\ldots\ldots\ldots (8)$$

If $\frac{4\omega^2 L^2}{R_L^2}$ "1, then

$$\therefore r = \frac{\sqrt{2}R_L}{3\,\omega L} = \frac{R_L}{1600L} \quad \ldots\ldots\ldots\ldots\ldots\ldots\ldots (9)$$

For $f = 60Hz$, L in henry and R_L in ohms

Equation (9) shows that the ripple will be reduced for a larger value of L. Ripple will increase if the load is reduced (R_L increased)

∴ Inductor Filter is used where R_L is very small, where the load current is high. (Heavy Load)

Output Voltage:

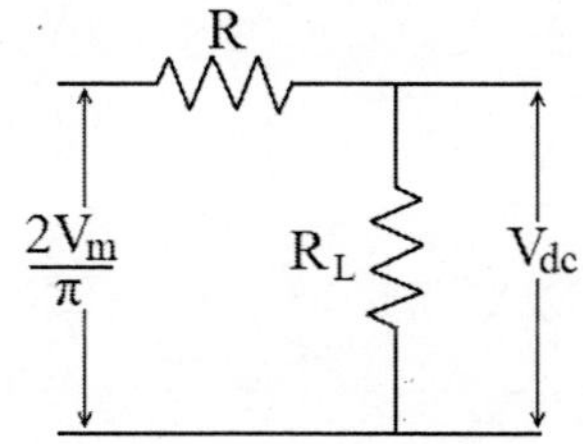

$$V_{dc} = \frac{2V_m}{\pi} \quad \ldots\ldots\ldots\ldots\ldots\ldots\ldots (10)$$

(For a perfect inductor with no resistance)

However, an inductive filter is used when R_L is low, the choke resistance is significant.

∴ The DC output voltage $V_{dc} = \frac{2V_m}{\pi} \cdot \frac{R_L}{R_L + R} \quad \ldots\ldots\ldots\ldots (11)$

Where R is the total resistance of choke.

$$\therefore V_{dc} = \frac{2V_m}{\pi} - I_{dc}R_L \quad \ldots\ldots\ldots\ldots\ldots (12)$$

Equation (12) shows that Inductor with high resistance causes poor voltage regulation.

$$\text{No Load Voltage} = \frac{2V_m}{\pi}$$

Experimental Determination of ripple factor

Suppose the output of a power supply is as shown below.

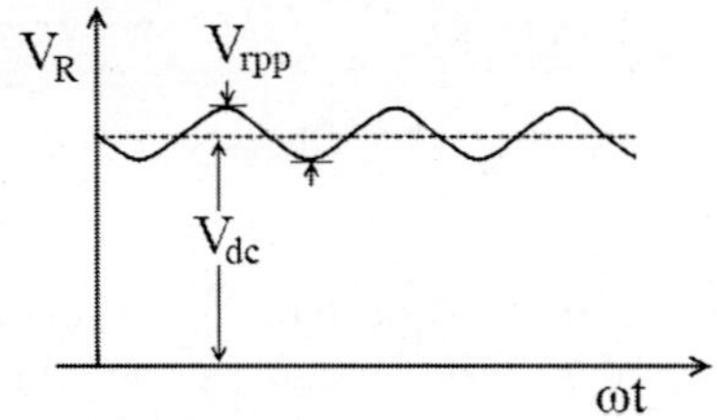

Fig. 1.27: Output waveform of an inductor filter

V_{rpp} = peak to peak value of ripple or AC component riding on DC level of V_{dc}

2.4.2. Capacitor Filter

This filter is used for light loads. The capacitor is directly connected across the load. Fig.1.28 shows the circuit diagram for a typical capacitor filter.

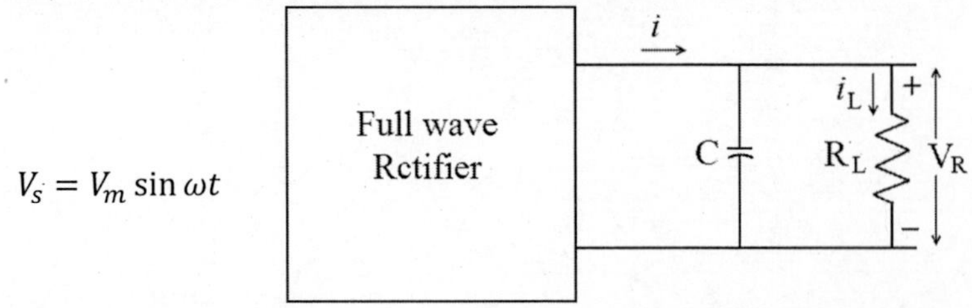

Fig. 1.28: Capacitor filter circuit

The capacitor charges up to the peak value of input voltage V_m and tries to maintain this value as the full-wave input drops to zero. The capacitor discharges through R_L until the input voltage increases to a value greater than the capacitor voltage. At this instant, the diode is forward-biased, and there will be a pulse of current through the diode, recharging the capacitor.

The output voltage remain V_m for light loads (large R_L). The discharge of C will be more significant, resulting in more ripple and lower DC output voltage as the load increases (R_L decreases).

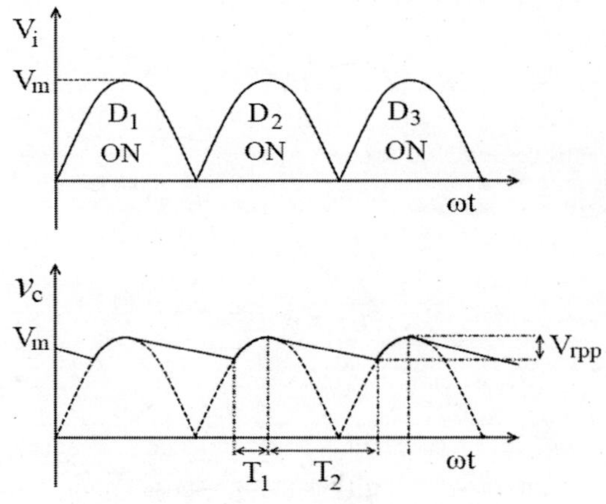

Fig. 1.29: Output of Capacitor filter

$$T = T_1 + T_2 = \frac{1}{f} = \frac{1}{2f_o}$$

$$T_1 + T_2 \cong T_2 = \frac{1}{2f_o}$$

f_o = Frequency of AC input of Rectifier.

Ripple Factor:

From Fig. 1.29, the average value of load current I_{dc} is the average value f_o the capacitor discharge current over an interval of T_2. The amount of charge lost by the capacitor during this interval (T_2),

$$\text{Q discharge} = I_{dc} \times T_2 \; \ldots\ldots\ldots\ldots\ldots (1)$$

The capacitor is again recharged by this value during the interval T_1. During T_1 the voltage across the capacitor changes by an amount equal to the peak-to-peak voltage of the ripple $V_{r(p-p)}$.

We know, $Q = VC$

$$\therefore \text{The change in charge, Q charge} = V_{rp-p} \times C \; \ldots\ldots\ldots (2)$$

$$\text{But Q charge} = \text{Q discharge} \; \ldots\ldots\ldots\ldots (3)$$

$$\therefore V_{r\,p-p} \times C = I_{dc} \times T_2 \; \ldots\ldots\ldots\ldots\ldots\ldots\ldots (4)$$

$$\therefore V_{r\,p-p} = \frac{I_{dc}}{C} \times T_2 \; \ldots\ldots\ldots\ldots\ldots\ldots (5)$$

Let us assume the load is light, the ripple is small, and the recharging time is smaller than the discharge time. i.e.

$$T_1 \| T_2 \;\; \& \;\; T_2 = \frac{T}{2} = \frac{1}{2f_o} \; \ldots\ldots\ldots\ldots\ldots\ldots (6)$$

$$Then, \;\; V_{r\,p-p} = \frac{I_{dc}}{2f_o C} \; \ldots\ldots\ldots\ldots\ldots\ldots\ldots (7)$$

$\therefore$ The ripple is triangular with peak to peak value of $V_{r\,p-p}$ and rms value of

$$V_r, rms = \frac{V_r, p-p}{2\sqrt{3}} \; \ldots\ldots\ldots\ldots\ldots (8)$$

$$\left(\text{since for triangular wave } rms = \frac{max}{\sqrt{3}}\right)$$

Now we know,

$$I_{dc} = \frac{V_{dc}}{R_L}$$

$$\therefore V_r, rms = \frac{I_{dc}}{4\sqrt{3} f_o C} = \frac{V_{dc}}{4\sqrt{3}\, f_o C\, R_L} \; \ldots\ldots\ldots\ldots\ldots (9)$$

Now Since $r = \frac{V_r, rms}{V_{dc}}$

$$\therefore \;\; r = \frac{1}{4\sqrt{3} f_o\, C\, R_L} \; \ldots\ldots\ldots\ldots (10)$$

$$\therefore \;\; r = \frac{24 \times 10^{-3}}{C R_L} \text{ if } f_o = 60Hz, R_L \; in \; \Omega \; \ldots\ldots\ldots (11)$$

Output voltage:

$$V_{dc} = V_m - \frac{V_{r,p-p}}{2} \; \ldots\ldots\ldots\ldots\ldots\ldots\ldots (13)$$

$$V_{dc} = V_m - \frac{I_{dc}}{4f_o C} \qquad \left[\because \; V_{r,p-p} = \frac{I_{dc}}{2f_o\, C} \; and \; I_{dc} = \frac{V_{dc}}{R_L}\right]$$

$$\therefore \; V_{dc} = \left(\frac{4f_o R_L\, C}{4f_o R_L\, C + 1}\right) V_m \; \ldots\ldots\ldots\ldots (14)$$

2.4.3. Zener Diode as a Voltage Regulator

The ability of the Zener diode to maintain a constant voltage on changing the current through it serves the purpose. Therefore, it can be used to get constant voltage from an unregulated power source. Fig.1.30 shows the circuit diagram for a Zener diode as a voltage regulator.

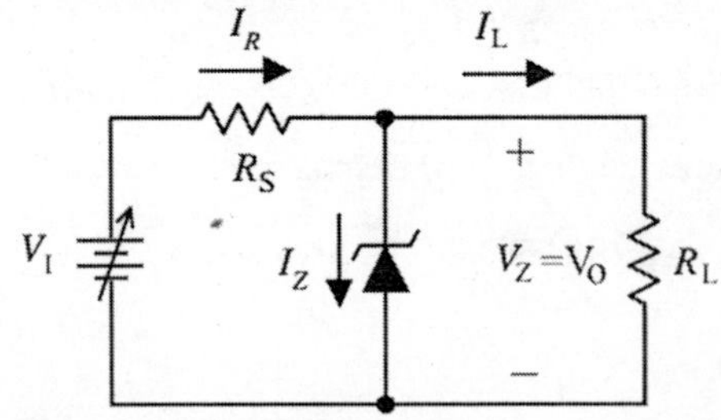

Fig.1.30: Zener diode as a voltage regulator

The unregulated power source, which provides the input voltage V_I, is connected in parallel to the Zener diode through a series resistance R_S R_L is the load resistor across which steady output is preferred and is connected in parallel to the Zener diode. The total current (I) passing through R_s is the sum of diode current (I_Z) and load current (I_L).

When the input voltage increases, the current through series resistance R_S and Zener diode increases. However, as in the breakdown region, the voltage of the Zener diode remains constant though the current may change. Hence the voltage drop across series resistance R_S increases; in turn, a constant voltage across the load is maintained. Similarly, when the input voltage decreases, the current through the Zener diode and R_S decreases.

It is noted that voltage drop decreases across R_S occur without any change in voltage across the Zener diode. As a result, the output voltage remains the same across the load. Hence, the load voltage is regulated, and the Zener diode acts as a voltage regulator.

The current flow series resistance Rs is given by

$$I_R = \frac{V_I - V_Z}{R_S}$$

The voltage drop across the Zener diode is

$$V_O = V_Z + I_Z R_Z$$

The current flow through load resistance is

$$I_L = \frac{V_O}{R_L}$$

Hence, $I_S = I_Z + I_L$

$$Iz(max) = \frac{V_{max} - V_z}{R_s} - I_{L(\text{min})}$$

$$Iz(min) = \frac{V_{min} - V_z}{R_s} - I_{L(\text{max})}$$

Based on the above relationships, the maximum and minimum current limiting resistance are given as follows:

$$R_{S(min)} = \frac{V_{(max)} - V_Z}{I_{Z(max)} + I_{L(min)}}$$

$$R_{S(max)} = \frac{V_{(min)} - V_Z}{I_{Z(min)} + I_{L(max)}}$$

For proper operation of the Zener diode shunt regulator $\boldsymbol{R_{S(min)} < R_S < R_{S(max)}}$

Disadvantages of Zener shunt regulators

1. High regulation factor
2. High output resistance
3. It has a high power dissipation in series resistance Rs and Zener diode compared to output power.

2.5. Breakdown Mechanisms

Under normal circumstances, significantly less current flows through a reverse-biased *P-N* junction diode. However, as the reverse voltage is increased gradually, a point is reached where the breakdown of the junction occurs due to an excessive increase in the current. Breakdown mechanisms in reverse biased diode are in two forms: (a) Zener breakdown and (b) Avalanche Breakdown.

2.5.1. Zener Breakdown: The Zener breakdown occurs in heavily doped *P-N* junctions, making the depletion layer extremely thin of the nanometer order. Due to heavy doping, the conduction band of *N-type* levels with the valence band of *P-type*

Thus large numbers of occupied states in the valence band of the P-side are brought to the same energy level as several unoccupied states in the conduction band of the *N-type*. As the barrier between them is very thin, quantum tunneling of electrons occurs with the supply of small reverse voltage. Thus electrons travel from the valence band of *P-type* to the conduction band of *N-type*, suddenly increasing the reverse current from *n* to *p* side.

Due to the drifting of electrons, there is a maximum current limit. This breakdown occurs at low reverse voltages of 6 V or less. The breakdown voltage for a particular diode decreases with an increase in temperature. Fig. 1.31 shows the band diagram of a Zener breakdown.

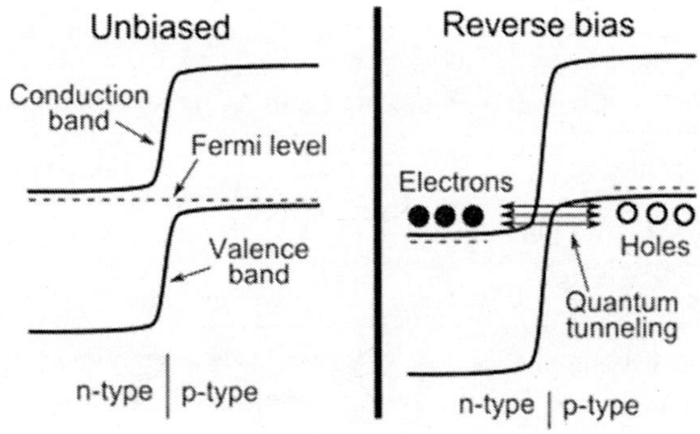

Fig. 1.31: Band diagram of Zener breakdown

2.5.2. Avalanche Breakdown: The diode with lesser doping undergoes avalanche breakdown when a high reverse voltage is applied. The lesser doping means the depletion width is significant, and hence electric field within the depletion region is not so high. Therefore the electric field will not be able to pull out electrons from the outer shell of atoms, and breakdown doesn't occur in the depletion region. However, due to a significant depletion region, the minority charge carriers moving through the depletion region get accelerated by the electric field.

Hence, minority charge carriers acquire high velocity and kinetic energy. When these charge carriers strike with atoms in the N-type and P-type regions, the high kinetic energy gets converted to thermal energy. Hence, electrons from the outermost shell are pulled out due to this energy, and a large current starts flowing. This type of breakdown is called avalanche breakdown. However, the temperature rises due to the high thermal energy, and the diode gets burned. As a result, the simple diodes (where avalanche breakdown occurs) are not commonly used. Instead, Zener diodes are preferred. Fig. 1.32 shows the *I-V* characteristics of Zener and Avalanche Breakdown.

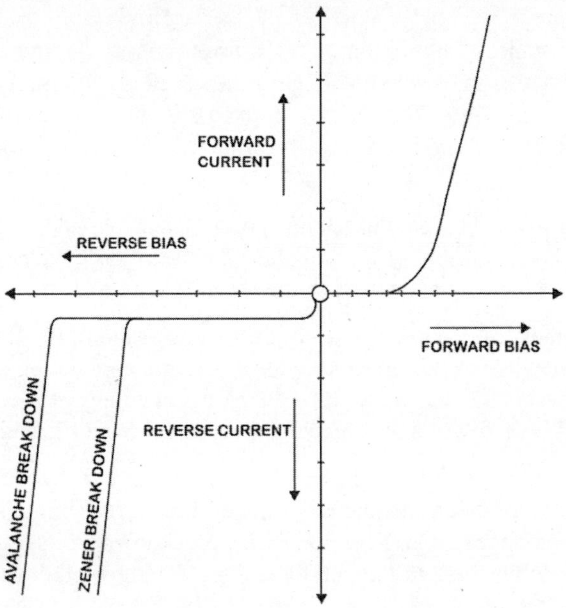

Fig. 1.32: *I-V* characteristics of Zener and Avalanche Breakdown.

Table:1.1: Comparision of Zener and Avalanche Breakdown

Zener Breakdown	Avalanche Breakdown
It occurs in a heavily doped *P-N* junction diode.	It occurs in a lightly doped *P-N* junction diode.
The depletion region is very thin (~10 nm).	The depletion region is comparatively wider.
Low reverse voltage must be applied across the junction (depending on the depletion layer's width).	Requires high reverse voltage to be applied across the junction.
The reverse field across the junction is such that it exerts a strong force on the bound electron, tearing it out from the covalent bond of the atoms.	The minority charge carriers (high K.E.) disrupt a covalent bond while colliding with host atoms creating electron-hole pair.

It occurs for breakdown voltages below 6 V.	It occurs at higher voltages.
It has a negative temperature coefficient of breakdown voltage.	It has a positive temperature coefficient of breakdown voltage.

As Zener breakdown voltage is less than that of avalanche breakdown voltage, Zener breakdown is said to occur before the avalanche breakdown. Hence, if doping of a diode is increased, the chances of Zener breakdown increases, and therefore breakdown voltage decreases.

Doping ↑ Breakdown voltage (V_B) ↓

Doping ↓ Breakdown voltage (V_B) ↑

2.6. Diode Equivalent Circuit

If the diode is not operating in the breakdown region, the diode can be modeled as a simple circuit element.

2.6.1. DC diode model

To define the DC diode model, the characteristics of an ideal diode need to be considered and the modifications required due to practical considerations. To revise:

- Ideal diode: $V_{ON} = 0$, $R_r = \infty$ and $R_f = 0$. In other words, the ideal diode is short in the forward bias region and open in the reverse bias region.
- Practical diode (silicon): $V_{ON} = 0.7V$, $R_r < \infty$ (typically several MΩ), $R_f \approx r_d$ (typically < 50 Ω).

The general representation for a practical diode under DC operating conditions are shown in Fig. 1.33. A diode is a two-terminal device. Hence, other circuit elements can be connected, as shown in Fig. 1.33. Here, the terminal voltage V_{ab} is the same as the voltage applied across the diode, v_D.

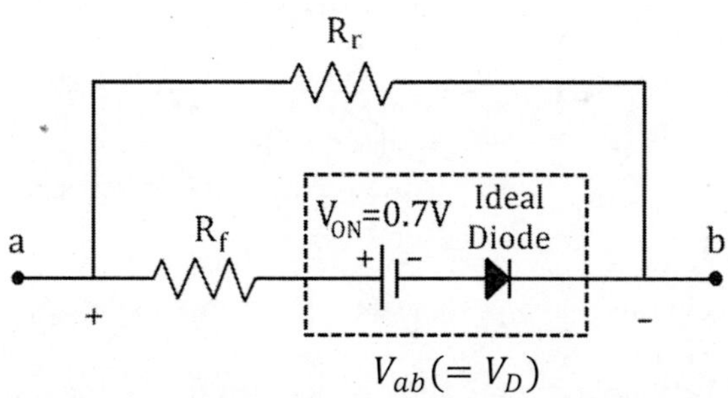

Fig. 1.33: A practical diode under DC operating conditions

For the forward bias region ($v_D \geq 0.7$ V for silicon), the ideal diode is short, and the terminal characteristics of the model above reduce to the parallel combination of R_r and R_f (Since $R_r >> R_f$, $R_r || R_f \approx R_f$).

Likewise, when the voltage applied to the diode is less than V_{ON} ($V_D < 0.7$V for silicon), the ideal diode is open, and the resistance between terminals *a* and *b* is R_r. These two cases are shown in Fig. 1.34

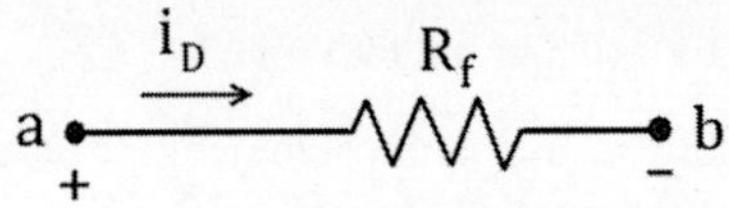

$V_{ab}(= V_D) > V_{ON}$

Fig. 1.34 (a): Forward biased DC diode model

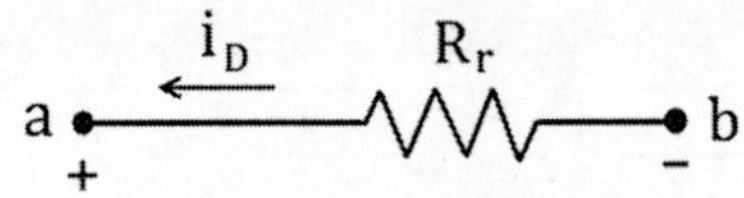

$V_{ab}(= V_D) < V_{ON}$

Fig. 1.34 (b): Reverse biased DC diode model

2.6.2. AC diode model

The diode model under AC conditions is complicated as a capacitive effect is generated whenever there is a charge separation. This charge separation of the diode is due to the depletion region, dependent on the applied bias. For the reverse bias condition, a junction capacitance (C_j) is introduced in parallel with the reverse bias resistance (R_r) as shown in Fig. 1.35.

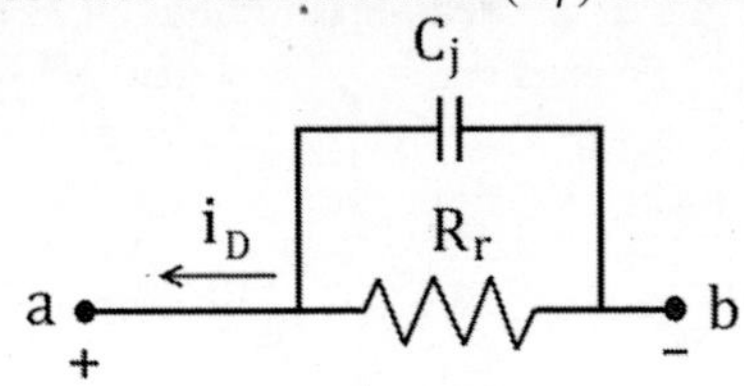

$V_{ab}(= V_D) < V_{ON}$

Fig. 1.35: Reverse biased ac diode model

Under AC operation, the charges are moving in the semiconductor material due to the current flow. However, the charges cannot move instantaneously. Hence, a charge storage effect leads to a diffusion capacitance (CD). Furthermore, the forward bias resistance is a function of the frequency. Therefore, the dynamic resistance, r_d replaces the constant R_f term. The AC diode model under forward bias conditions is illustrated in Fig. 1.36.

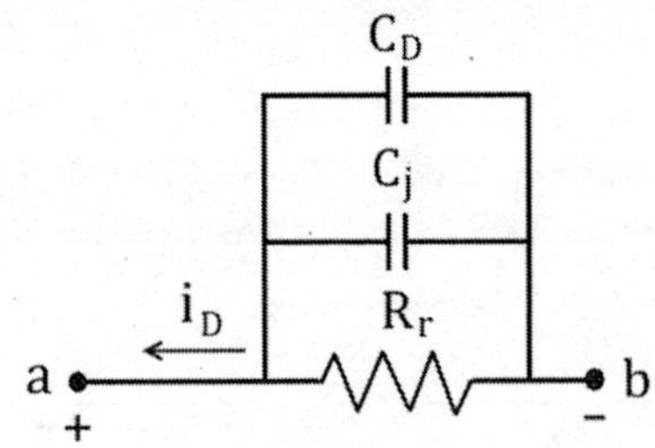

$V_{ab}(= V_D) > V_{ON}$

Fig. 1.36: Forward biased ac diode model

2.6.3. diode as a switch

Diodes are also used in circuits that mix signals together (mixers), detect the presence of a signal (detector), and act as a switch to open or close a circuit. Diodes used in such applications are called signal diodes. The simplest application of a signal diode is a diode switch, as shown in Fig. 1.37.

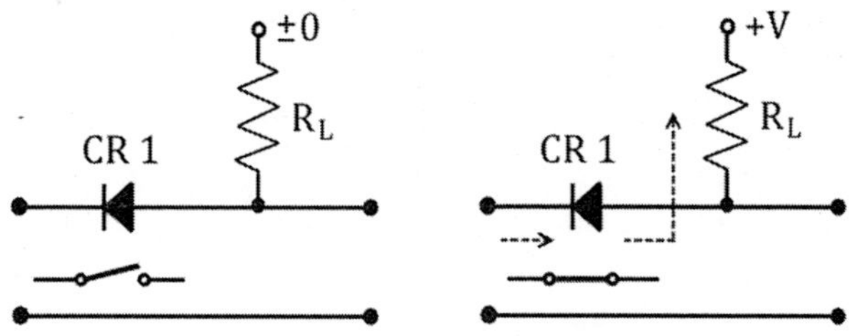

Fig. 1. 37: Diode as a switch

The diode is forward-biased when the input to this circuit is at zero potential. This is due to the zero potential on the cathode and the positive voltage on the anode. In this condition, the diode conducts and acts as a straight piece of the wire due to its very low forward resistance. As a result, the input is directly coupled to the output resulting in zero volts across the output terminals. Therefore, the diode acts as a closed switch when its anode is positive to its cathode.

If a positive input voltage is applied (equal to or greater than the positive voltage supplied to the anode) to the diode's cathode, the diode will be reverse-biased. In this situation, the diode is cut-off mode and acts as an open switch between the input and output terminals. Consequently, with no current flow in the circuit, the positive voltage on the diode's anode will be observed at the output terminal. Therefore, the diode acts as an open switch when reverse-biased.

3. Special Purpose Diodes

3.1. Light Emitting Diode (LED):

Light emitting diode is a *P-N* junction that, under appropriate forward biased circumstances, can serve as a light emitter. LED can emit spontaneous radiations in the electromagnetic spectrum's ultraviolet, visible, and infrared regions. An eye is only sensitive to light of energy $h\vartheta \geq 1.6\ eV$. This requires that the semiconductor used for LED. should have energy band gap larger than 1.6 *eV*. The LEDs fabricated on Gallium phosphide (GaP) of $E_g = 2.25\ eV$ substrates emit visible light.

Working Principle: Fig. 1.38 shows the symbol and circuit diagram of a LED. The LED is a semiconductor device based on the principle of electro-luminescence. It emits light in forward biased condition.

When energy is supplied to the semiconductor material the electrons are excited to higher energy state E_2 and after 10^{-8} *seconds* (mean life time) ultimately spontaneously fall back to E_1 state with an emission of photon of energy $E = E_2 - E_1$. LED is mainly concerned with injection electroluminescence i.e. since it is forward biased electrons are injected into the *P-side* and holes are injected to *N-side* by the battery. Since in P-side holes are the majority charge carriers, injected electrons are minority charge carriers and in *n*-region electrons are the majority carriers obviously the injected holes are minority carriers. These minority charge carriers when recombine with the majority charge carriers in the respective regions emit energy in the form of photon of light.

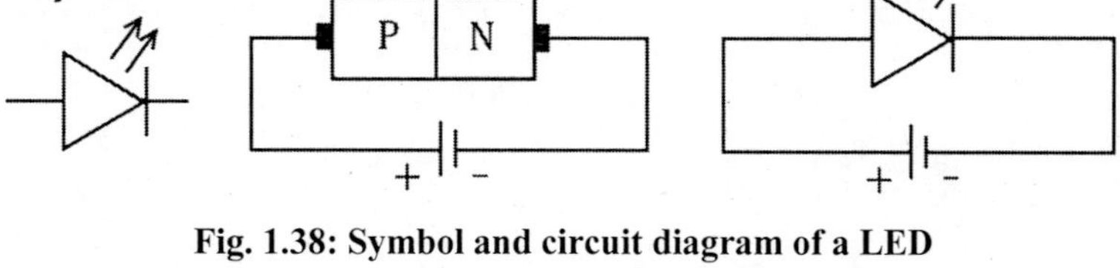

Fig. 1.38: Symbol and circuit diagram of a LED

3.2. Zener Diode:

Zener diode is a *P-N* junction diode connected in reverse bias condition. It is mainly operated in the breakdown region of reverse bias. By varying the concentration of the dopants and other factors, the breakdown voltage can be designed for specific applications. Zener breakdown occurs when a *P-N* junction connected to a high reverse bias field, such that bound electrons are torn apart from the covalent bonds. This creates a large number of electron-hole pairs thus increasing the reverse current.

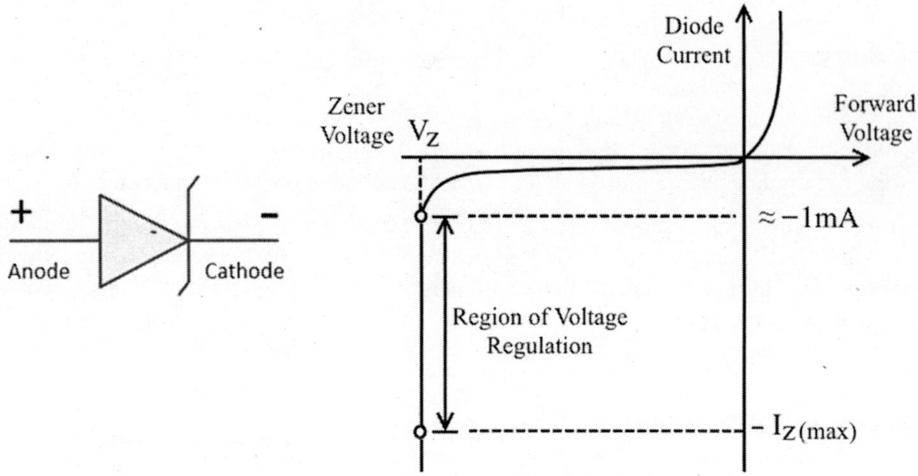

Fig. 1.39: Symbol and *I-V* characteristics of a Zener diode.

Fig. 1.39 shows the symbol of a Zener diode and its *I-V* characteristics. As the reverse voltage is increased, the reverse current remains constant till a certain value beyond which it increases abruptly. The voltage at which the reverse current suddenly increases is called Zener voltage. Zener diodes are operated at low voltages less than 6V. Contrary to ordinary diodes, the breakdown phenomenon in Zener diodes is reversible and does not damage the diode. From Fig.1.39 it can be deduced that in the Zener region, the voltage across it is constant though the current changes depending on the supply voltage. Due to such inexplicable properties Zener diodes are employed in voltage regulators, reference diode, etc.

3.2.1. Characteristics of Zener diodes

- **Zener Voltage (Vz):** The zener voltage depends upon the type of zener diode used. It varies from 1.8 V to 2000V.
- **Maximum Power Dissipation (Pz max):** the power dissipation of zener diode is the product of zener voltage and its maximum current.

  $P_{Z\,MAX} = V_Z \times I_{Z\,MAX}$
- **Zener current ($I_{Z\ M.A.X.}$):** it is the ratio of power dissipation to the zener voltage. Generally it varies from 150mW to 50W.

  $P_{Z\,MAX} = V_Z \times I_{Z\,M.A.X.}$
- **Zener Resistance (Rz):** It varies from few ohms to 50 ohm. For most commonly used zener diode 1N4370 , the resistance is 30ohm.

3.2.2. Application of Zener diode:

- Voltage regulator

- Over voltage protection circuits
- As a limiters in wave shaping circuits
- In transistor biasing provides a fixed reference voltage.

3.3. Photo Diode:

The diagrams shown below are the Construction, biasing, and symbol of Photo diode. The construction details of a typical photodiode is as shown in Fig.1.40.

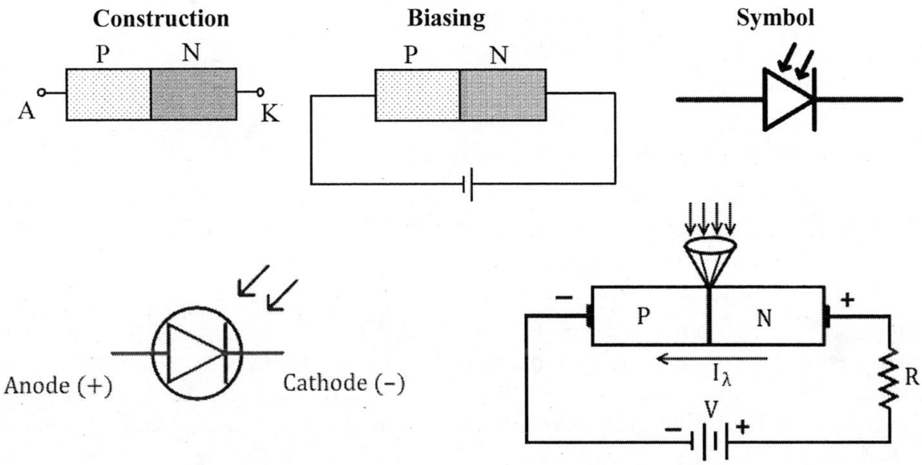

Fig. 1.40: The details of a photo diode

With light falls on reverse biased P-N photo junction, holes and electron pairs are liberated which leads to current flow through the external load. Current will be zero only for a positive voltage V_T

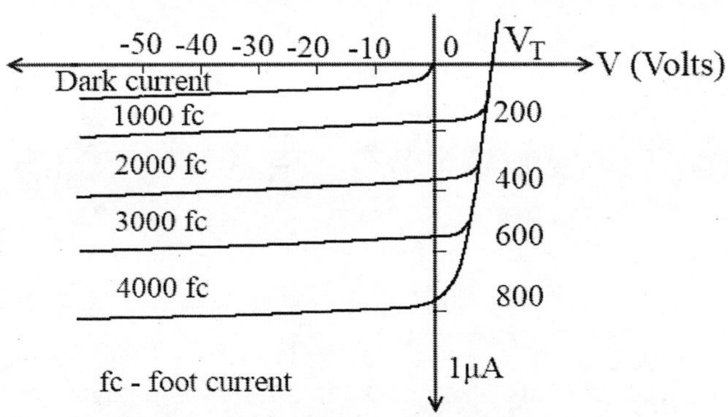

Fig. 1.41: V-I characteristics of a photo diode

A photodiode is a type of photodetector capable of converting light into either current or voltage, depending upon the mode of operation. A photodiode is a P-N junction or P.I.N. structure. It is designed to operate in reverse bias. When a photon of sufficient energy strikes the diode, it excites an electron, creating a *free electron* (and a positively charged electron *hole*). This mechanism is also known as the inner photoelectric effect.

If the absorption occurs in the junction's depletion region or one diffusion length away, these carriers are swept from the junction by the built-in field of the depletion region. Thus, holes move toward the anode, and electrons toward the cathode, producing a photocurrent. This photocurrent is the sum of the dark current (without light) and the light current, so the dark current must be minimized to enhance the device's sensitivity.

3.3.1. Applications

- Camera: Light Meters, Automatic Shutter Control
- Medical : CAT Scanners, X ray Detection. Pulse Oximeters.
- Safety Equipment: Smoke Detectors, Flame Monitors.
- Automotive: Headlight Dimmers.
- Communications: Fiber Optic Links
- Industry: Bar Code Scanners.

3.4. Silicon Controlled Rectifiers

Within the family of ***pnpn*** devices, the silicon-controlled rectifier (SCR) was first introduced in 1956 by Bell Telephone Laboratories.

As the terminology indicates, the SCR is a rectifier constructed of silicon material with a third terminal for control purposes. Silicon was chosen because of its high temperature and power capabilities. The basic operation of the SCR is different from the fundamental two-layer semiconductor diode in that a third terminal, called a gate, determines when the rectifier switches from the open-circuit to short-circuit state. It is not enough to simply forward-bias the anode-to-cathode region of the device. In the conduction region, the dynamic resistance of the SCR is typically 0.01 to 0.1 Ω. The reverse resistance is typically 100 kΩ or more. The graphic symbol for the SCR is shown in Fig. 1.42 with the corresponding connections to the four-layer semiconductor structure.

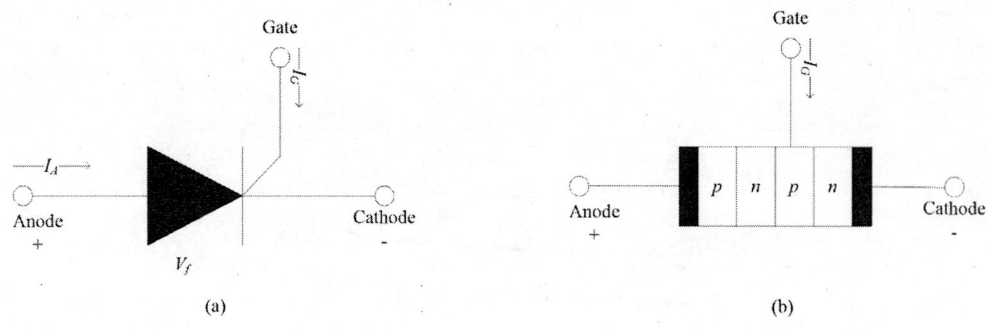

Fig. 1.42. (a) SCR symbol; (b) basic construction.

As indicated in Fig. 1.42(a), if forward conduction is to be established, the anode must be positive with respect to the cathode. This is not, however, a sufficient criterion for turning the device on. A pulse of sufficient magnitude must also be applied to the gate to establish a turn-on gate current, represented symbolically by I_{GT}. Fig. 1.43 shows the two transistor terminology of a SCR. Here, one transistor is an *npn* device while the other is a *pnp* transistor.

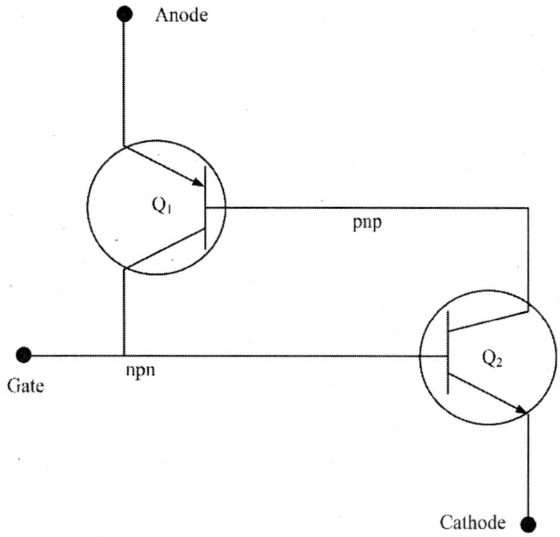

Fig. 1.43. Two transistor terminology of a SCR

Let the signal shown in Fig. 1.44(a) will be applied to the gate of the circuit of Fig. 1.44(b). During the interval $0 \rightarrow t_1$, $V_{gate} = 0$ V, the circuit of Fig. 1.43 will appear as shown in Fig. 1.44(b) ($V_{gate} = 0$ V is equivalent to the gate terminal being grounded as shown in the figure). For $V_{BE_2} = V_{gate} = 0$ V, the base current $I_{B_2} = 0$ and I_{C_2} will be approximately I_{CO}. The base current of Q_1, $I_{B_1} = I_{C_2} = I_{CO}$, is too small to turn Q_1 on. Both transistors are therefore in the "off" state, resulting in a high impedance between the collector and emitter of each transistor and the open-circuit representation for the controlled rectifier as shown in Fig. 1.44(c).

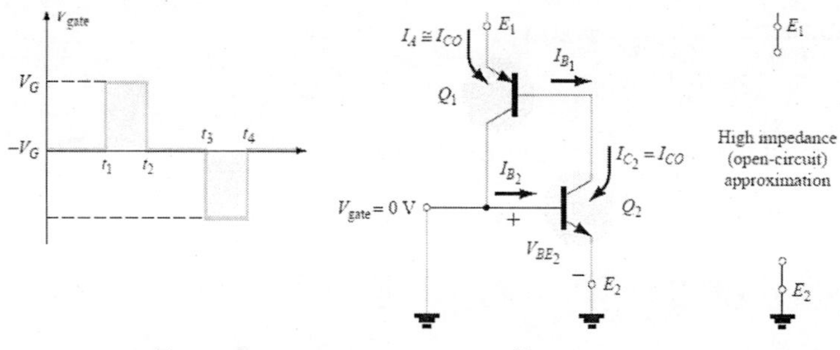

Fig. 1.44. "Off" state of the SCR

At $t = t_1$, a pulse of V_G volts will appear at the SCR gate. The circuit conditions established with this input are shown in Fig. 1.45(a). The potential V_G was chosen sufficiently large to turn Q_2 on ($V_{BE_2} = V_G$). The collector current of Q_2 will then rise to a value sufficiently large to turn Q_1 on ($I_{B_1} = I_{C_2}$). As Q_1 turns on, I_{C_1} will increase, resulting in a corresponding increase in I_{B_2}. The increase in base current for Q_2 will result in a further increase in I_{C_2}. The net result is a regenerative increase in the collector current of each transistor. The resulting anode-to-cathode resistance ($R_{SCR} = V/I_A$) is then small because I_A is large, resulting in the short-circuit representation for the SCR as indicated in Fig. 1.45(b). The regenerative action described above results in SCRs having typical turn-on times of 0.1 to 1μs. However, high-power devices in the range 100 to 400 A may have 10 to 25μs turn-on times.

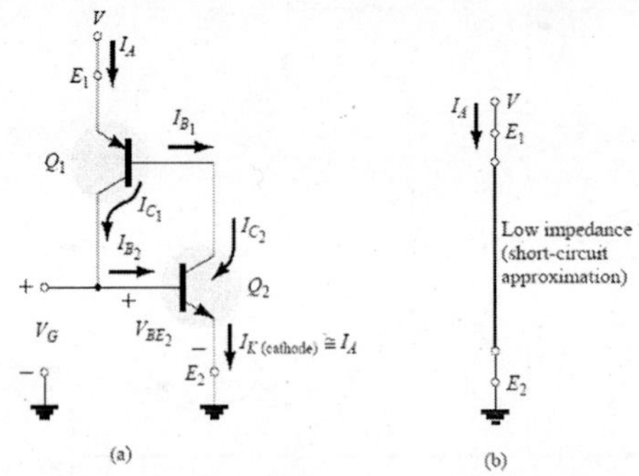

Fig. 1.45. "On" state of the SCR

The next question of concern is how is turnoff accomplished? An SCR *cannot* be turned off by simply removing the gate signal, and only a special few can be turned off by applying a negative pulse to the gate terminal as shown in Fig. 1.44(a) at $t = t_3$.

Forced commutation is the "forcing" of current through the SCR in the direction opposite to forward conduction. In forced communication, the turn-off circuit consists of an *npn* transistor, a dc battery V_B, and a pulse generator. During SCR conduction, the transistor is in the "off" state, that is, $I_B = 0$ and the collector-to-emitter impedance is very high (for all practical purposes an open circuit). This high impedance will isolate the turn-off circuitry from affecting the operation of the SCR. For turn-off conditions, a positive pulse is applied to the base of the transistor, turning it heavily on, resulting in a very low impedance from collector to emitter (short-circuit representation). The battery potential will then appear directly across the SCR, forcing current through it in the reverse direction for turn-off. Turn-off times of SCRs are typically 5 to 30μs.

3.4.1 SCR Characteristics and Ratings

The characteristics of an SCR are provided in Fig. 1.46 for various values of gate current. The currents and voltages of usual interest are indicated on the characteristic.

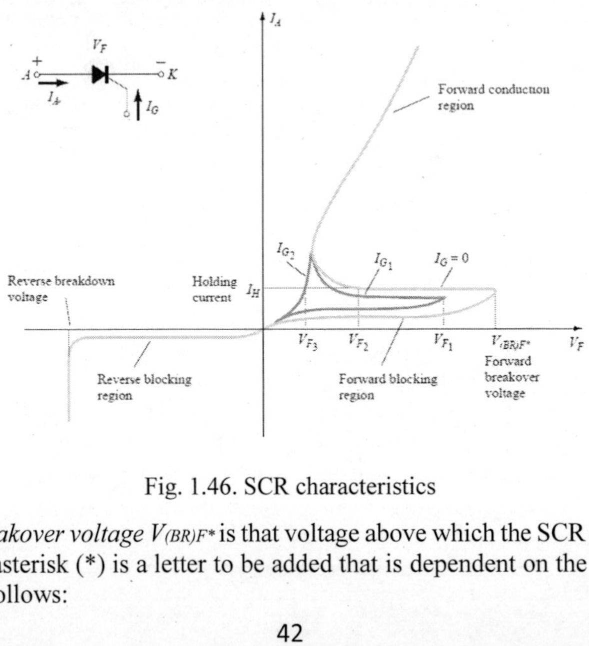

Fig. 1.46. SCR characteristics

- *Forward breakover voltage* $V_{(BR)F^*}$ is that voltage above which the SCR enters the conduction region. The asterisk (*) is a letter to be added that is dependent on the condition of the gate terminal as follows:

O= open circuit from *G* to *K*
S= short circuit from *G* to *K*
R= resistor from *G* to *K*
V= fixed bias (voltage) from *G* to *K*

- *Holding current* (I_H) is that value of current below which the SCR switches from the conduction state to the forward blocking region under stated conditions.
- *Forward and reverse blocking regions* are the regions corresponding to the opencircuit condition for the controlled rectifier which *block* the flow of charge (current) from anode to cathode.
- *Reverse breakdown voltage* is equivalent to the Zener or avalanche region of the fundamental two-layer semiconductor diode.

Applications: The more common areas of application for SCRs include relay controls, time-delay circuits, regulated power suppliers, static switches, motor controls, choppers, inverters, cycloconverters, battery chargers, protective circuits, heater controls, and phase controls.

Solved Examples

Ex. 1. Find the built-in voltage for a *Si* P-N junction with $N_A = 10^{15} cm^{-3}$ and $N_D = 10^{17} cm^{-3}$

Solution: Assume $n_i = 10^{10} cm^{-3}$;

$$V_{bi} = \frac{kT}{q} . \ln \left(\frac{N_D N_A}{n_i^2} \right)$$

Important values to remember

At room temperature, $T \approx 300K$,

$$kT \approx 0.026eV$$

$$\left(\frac{kT}{q} \right) = 0.026V$$

$$\therefore V_{bi} = 0.026 \times \ln \left(\frac{10^{17} \times 10^{15}}{10^{10}} \right)$$

$$\therefore V_{bi} = 0.718V$$

Ex. 2. If V_{rpp} is 20 V and V_{dc} is 200 V. Find ripple factor (% ripple r).

Solution:

$$V_r, max = \frac{V_{rpp}}{2} = \frac{20}{2} = 10V$$

$$V_r, rms = \frac{V_r, max}{\sqrt{2}} = \frac{10}{\sqrt{2}} = 7.10V$$

$$\text{Ripple Factor } (r) = \frac{V_r, rms}{V_{dc}} = \frac{7.10}{200} = 0.0355$$

$$\therefore \% \text{ Ripple} = r \times 100 = 0.0355 \times 100$$

$$= 3.55\%$$

Ex. 3. Determine the DC resistance levels for the curve shown below-

(a) $I_D = 2$ mA

(b) $I_D = 20$ mA

(c) $V_D = -10$ V

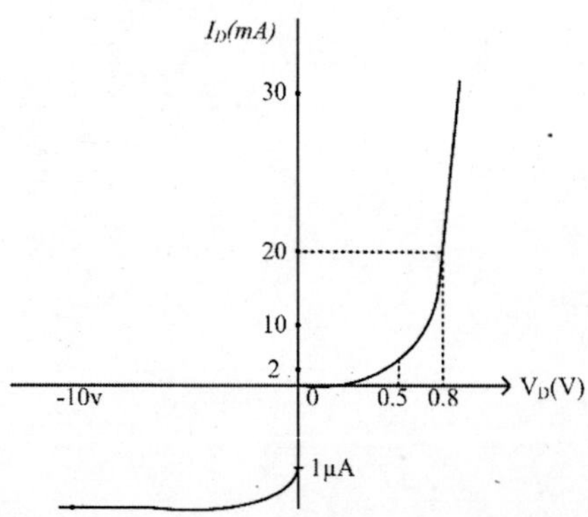

Solution:

(a) At $I_D = 2$ mA, $V_D = 0.5$ V (from the curve) and

$$R_D = \frac{V_D}{I_D} = \frac{0.5V}{2mA} = \mathbf{250\Omega}\ [\boldsymbol{Ans.}]$$

(b) At $I_D = 20$ mA, $V_D = 0.8$ V (from the curve) and

$$R_D = \frac{V_D}{I_D} = \frac{0.8V}{20mA} = \mathbf{40\Omega}\ [\boldsymbol{Ans.}]$$

(c) At $V_D = -10$ V, $I_D = -I_s = -1\mu$ A (from the curve) and

$$R_D = \frac{V_D}{I_D} = \frac{10\ V}{1\mu A} = \mathbf{10M\Omega}\ [\boldsymbol{Ans.}]$$

Exercise Questions

1. Can an ordinary rectifier diode be used as a zener diode? Discuss and explain.
2. Define the term transition capacitance C_T of P-N junction diode.
3. Discuss the ideal diode current–voltage relationship. Describe the meaning of I_O and V_T.
4. Distinguish between avalanche and zener mechanisms.
5. Explain the effect of the temperature of a diode.
6. The reverse saturation current I_O in a germanium diode is 6 μA. Calculate the current flowing through the diode when the applied bias voltage is 0.2, 0.3, and 0.4V at room temperature. [Ans:13.15mA, 21mA, 28.8mA]
7. Explain the term diffusion capacitance C_D of a forward-biased diode.
8. Define a load line in a simple diode circuit.
9. Determine the forward bias voltage applied to a silicon diode to cause a forward current of 10 mA and reverse saturation current, $I_O = 25 \times 10^{-7}$ A at room temperature. [Ans:0.4V]
10. Discuss the working principle of a LED. State its applications.
11. Why is there a sudden increase in current in the Zener diode?
12. A half-wave rectifier has an input voltage of 240 V r.m.s. If the step-down transformer has a turn ratio of 8:1, what is the peak load voltage? Ignore the diode drop.[Ans:42.5 V]
13. Discuss the working of a P-N junction diode as a switch.
14. With neat sketches, explain the breakdown mechanism in a P-N junction diode.
15. With suitable diagrams and equations, explain the working of a Zener diode as a regulator.
16. Define the term rectifier. List the classification of a rectifier. Explain the construction and working principle of each.
17. Compare the halfwave, center-tapped full wave, and bridge rectifier.
18. Explain the construction and working of a photodiode.
19. Discuss the construction and working principle of an SCR.

Self Study Questions

1. Describe in your own words the conditions established by forward- and reverse-bias conditions on a *p-n* junction diode and how the resulting current is affected.
2. Describe how you will remember the forward- and reverse-bias states of the *p-n* junction diode. That is, how you will remember which potential (positive or negative) is applied to which terminal?
3. list three materials that have a negative temperature coefficient and three that have a positive temperature coefficient.
4. Sketch the atomic structure of copper and discuss why it is a good conductor and how its structure is different from germanium and silicon.
5. What is the important difference between the characteristics of a simple switch and those of an ideal diode?

Multiple Choise Questions

1. A crystal diode has forward resistance of the order of ....

a) kΩ
b) Ω
c) MΩ
d) none of the above

2. If the arrow of the crystal diode symbol is positive concerning the bar, then the diode is .... biased.

a) forward
b) reverse
c) either forward or reverse
d) none of the above

3. The reverse current in a diode is of the order of ....

a) kA
b) mA
c) μA
d) A

4. The forward voltage drop across a silicon diode is about ....

a) 2.5 V
b) 3 V
c) 10 V
d) 0.7 V

5. The DC resistance of a crystal diode is .... its AC resistance

a) the same as
b) more than
c) less than
d) none of the above

6. The leakage current in a crystal diode is due to ....

a) minority carriers
b) majority carriers
c) junction capacitance

d) none of the above

7. If the doping level of a crystal diode is increased, the breakdown voltage............
   a) remains the same
   b) is increased
   c) is decreased
   d) none of the above
8. If the temperature of a crystal diode increases, then leakage current ....
   a) remains the same
   b) decreases
   c) increases
   d) becomes zero
9. The knee voltage of a crystal diode is approximately equal to ....
   a) applied voltage
   b) breakdown voltage
   c) forward voltage
   d) barrier potential

10 A P-N junction diode is a .... device
   a) non-linear
   b) bilateral
   c) linear
   d) none of the above

11. When a crystal diode is used as a rectifier, the most important consideration is ....
   a) forward characteristic
   b) doping level
   c) reverse characteristic
   d) PIC rating

12. A Zener diode is used as ....
   a) an amplifier
   b) a voltage regulator
   c) a rectifier
   d) a multivibrator

13 The doping level in a zener diode is .... that of a crystal diode
   a) the same as
   b) less than
   c) more than
   d) none of the above

14 A zener diode is always .... connected.
   a) reverse
   b) forward
   c) either reverse or forward
   d) none of the above

15. The disadvantage of a half-wave rectifier is that the.................
   a) components are expensive
   b) diodes must have a higher power rating
   c) output is difficult to filter
   d) none of the above

16. If the AC input to a half-wave rectifier is an r.m.s value of $400/\sqrt{2}$ volts, then the diode PIV rating is ....

a) $400/\sqrt{2}$ V
b) 400 V
c) $400 \times \sqrt{2}$ V
d) none of the above

17. The ripple factor of a half-wave rectifier is ....
a) 2
b) 1.21
c) 2.5
d) 0.48

18. There is a need for a transformer for ....
a) half-wave rectifier
b) center-tap full-wave rectifier
c) bridge full-wave rectifier
d) none of the above

19. The PIV rating of each diode in a bridge rectifier is ....that of the equivalent center-tap rectifier
a) one-half
b) the same as
c) twice
d) four times

20 For the same secondary voltage, the output voltage from a center-tap rectifier is .... than that of a bridge rectifier
a) twice
b) thrice
c) four-time
d) one-half

21. The .... filter circuit results in the best voltage regulation
a) choke input
b) capacitor input
c) resistance input
d) none of the above

22 The maximum efficiency of a half-wave rectifier is ....
a) 40.6 %
b) 81.2 %
c) 50 %
d) 25 %

23 The most widely used rectifier is ....
a) half-wave rectifier
b) center-tap full-wave rectifier
c) bridge full-wave rectifier
d) none of the above

24. The color of emitted light from LED depends on
a) Construction of LED, that is physical dimensions
b) Number of available carriers
c) Type of semiconductor material used
d) Number of recombinations taking place

25 The typical value of power consumption of LED is
a) Around 10 mW
b) In between 15 mW and 20 mW
c) In between 30 mW and 40 mW

d) In between 35 mW and 50 mW

26 A photodiode is used to detect

a) Visible light
b) Invisible light
c) No light
d) Both visible and invisible light

27 When there is no incident light, the reverse current in a photodiode is essentially non-existent and is referred to as

a) Zener current
b) Dark current
c) Photocurrent
d) PIN current

28 The photocurrent is directly proportional to

a) forward current
b) reverse current
c) intensity of light
d) none of these

29 If the gate current of an SCR is increased, the forward breakdown voltage will….

a) Increase
b) Decrease
c) Not affect
d) Become infinite

30 After firing an SCR, the gate pulse is removed. The current in the SCR will…

a) Remain the same
b) Immediately fall to zero
c) Rise up
d) Rise a little and then fall to zero

MCQ Answer key:

1	**b**	11	**d**	21	**a**
2	**a**	12	**b**	22	**a**
3	**c**	13	**c**	23	**c**
4	**d**	14	**a**	24	**c**
5	**c**	15	**c**	25	**b**
6	**a**	16	**b**	26	**d**
7	**c**	17	**a**	27	**b**
8	**c**	18	**b**	28	**c**
9	**d**	19	**a**	29	**c**
10	**d**	20	**d**	30	**a**

ATTAINMENT & GAP ANALYSIS

Attainment of the Programme Outcomes will be compiled in the table below to make a Gap Analysis and work out remedial measures:

Course outcome	Attainment of the Programme Outcomes (1- Weak Correlation; 2- Medium correlation; 3- Strong Correlation)											
	PO-1	PO-2	PO-3	PO-4	PO-5	PO-6	PO-7	PO-8	PO-9	PO-10	PO-11	PO-12
CO-1												
CO-2												
CO-3												

@@@@@@@@@

2. Transistor Characteristics

RATIONALE

It is now more than 70 years since the first transistor was introduced to this world. Since then, it becomes an integral part of the majority of IC designs and modern switching mechanisms.

In this chapter, we will study the details of a semiconductor diode, construction and working principles of a few advanced and application-oriented diodes.

UNIT OUTCOMES

U2-O1: Unit-1 Learning Outcome-1

To know about the construction and basic operation of a transistor and its amplifying action.

U2-O2: Unit-1 Learning Outcome-2

To know about the operational characteristics of a transistor and different types of biasing mechanisms.

U2-O3: Unit-1 Learning Outcome-3

To know the construction and operation of a FET and its types and CMOS circuits.

LEARNING OBJECTIVES

LO1: Study of fundamentals of a transistor and its operation.

LO2: Study of Transistor as an amplifier, different configurations of the transistors.

LO3: Study of the concept of operating point and voltage divider bias circuit.

LO4: Study of construction and operation of FETs and their types.

LO5: Introduction to CMOS circuits

MAPPING THE UNIT OUTCOMES WITH THE COURSE OUTCOMES

Unit Outcome	EXPECTED MAPPING WITH COURSE OUTCOMES (1- Weak Correlation; 2- Medium correlation; 3- Strong Correlation)						
	CO-1	**CO-2**	**CO-3**	**CO-4**	**CO-5**	**CO-6**	**CO-7**
U2-O1	**2**	**1**	**3**	--	--	--	--
U2-O2	**2**	**1**	**3**	--	--	--	--
U2-O3	**2**	**1**	**3**	--	--	--	--

Interesting Facts:

4. **Bell Labs** publicly announced the first transistor at a press conference in New York on June 30, 1948. The transistor went on to replace bulky vacuum tubes and mechanical relays. The invention revolutionized the world of electronics and became the basic building block upon which all modern computer technology rests.
5. A transistor is like a miniature on-off switch that allows a computer to process information. A computer can't operate without an integrated circuit (chip), and a chip can't operate without a transistor. Transistors have shrunk in size with a factor of 222 since the first Intel 4004 chip was introduced in 1971
6. A transistor is an abbreviated combination of the words **"transfer", and "varistor".** The device logically belongs in the varistor family and has the transconductance or transfer impedance of a device having gain, so this combination is descriptive.
7. The first transistor was about the size of the **palm of a hand**, with a depth of two matchbooks stacked on top of each other. **The first commercial device to use a transistor was the Sonotone 1010 hearing aid, created in 1953.**
8. **As of 2022,** the largest transistor count in a commercially available microprocessor is **114 billion transistors, in Apple's ARM-based dual-die M1 Ultra system** on a chip, which is fabricated using TSMC's 5 nm semiconductor manufacturing process.
9. The first FET device to be successfully built was the junction field-effect transistor (JFET). A JFET was first patented by Heinrich Welker in 1945.
10. The invention of the MOSFET is credited to Mohamed Atalla and Dawon Kahng when they successfully fabricated the first working sample at Bell Labs in November 1959.
11. The MOSFET is by far the most common transistor and the basic building block of most modern electronics. **The MOSFET accounts for 99.9% of all transistors in the world.**
12. **Frank Wanlass** invented the complementary metal oxide semiconductor (CMOS), the technology employed in most modern microchips, at Fairchild Semiconductor, USA in 1963.

Video Resources:

Sr.	Title	URL	QR Code
1.	Transistor Working	https://www.youtube.com/watch?v=J4oO7PT_nzQ	
2.	Transistor Characteristics	https://www.youtube.com/watch?v=3jbTgOSRUGQ	
3.	Transistor Configuration	https://www.youtube.com/watch?v=yXjKrlvZd1o	
4.	Transistor Biasing and Q-Point	https://www.youtube.com/watch?v=5T84Jzcgj7M	
5.	Transistor as a Switch	https://www.youtube.com/watch?v=UIEGKvCfDOA	
6.	Transistor as an Amplifier	https://www.youtube.com/watch?v=4U_dwuRu6SA	

7.	Construction and working of FET	https://www.youtube.com/watch?v=_DZ7baOhNFQ	
8.	Working of a MOSFET	https://www.youtube.com/watch?v=NqnWcv3KXSA	
9.	Construction and working of Enhancement type MOSFET	https://www.youtube.com/watch?v=4_nGFY7zgDM	
10	Construction and working of Depletion type MOSFET	https://www.youtube.com/watch?v=XqGBNyhImV4	

2.1 Bipolar Junction Transistor (BJT)

Transistor means trans-resistor i.e. transfer of resistor. It is a device in which the same amount of current is transferred from low resistance region to high resistance region. A transistor is a sandwich of one type of semiconductor material (***p*** or ***n***) between two layers of the other type. Fig 2.1 shows the construction of PNP and NPN transistors respectively.

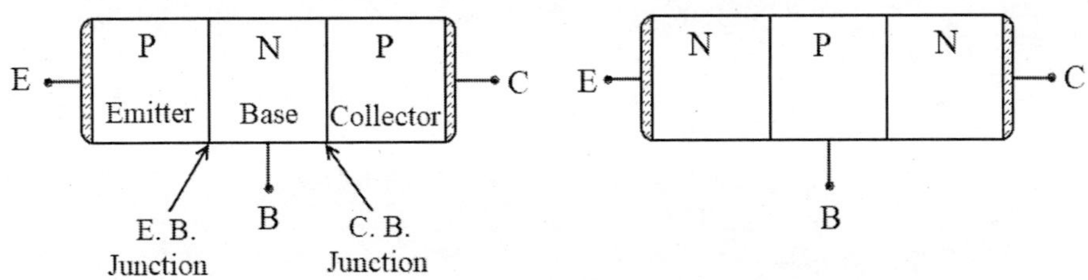

Fig. 2.1: Construction of BJT- PNP & NPN

Transistor consists of three layers called Emitter, Base & Collector. Each layer has terminals labeled E, B, and C respectively.

(a) **Emitter:** It generates large numbers of charge carriers for the conduction process. As it is a source of charge carriers it should be rich in the number of carriers and so it is **heavily doped.** It is **moderate in area**.

(b) **Base:** It controls the motion of charge carriers from emitter to collector. As the base is of the opposite type of emitter, many of the charge carriers from the emitter may be recombining in the base. Thus to avoid (minimizing) recombination in the base, it is **lightly doped**. It is **narrower** than E & C so charge carriers should not spend more time in the base.

(c) **Collector:** It collects charge carriers emitted by the emitter. It is **moderately doped.** Due to the collection of charge carriers, maximum power dissipation is in the collector region and so to withstand this high power dissipation, it must have **maximum area** compared to E &B.

Symbols

1. NPN

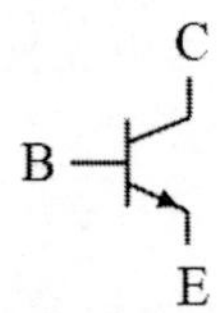

Electrons are the majority of carriers

2. PNP

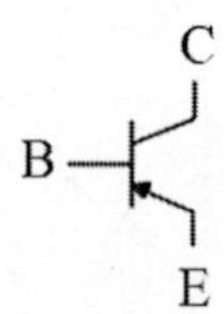

Holes are the majority of carriers

Arrowhead shows the direction of conventional current from p to n-Type. Terminal with arrow identifies the Emitter.

2.2 Modes (Regions) of operation of BJT

There are three modes

(a) **Cutoff:** Here both the junctions are reverse biased and so the output current is only due to minority charge carriers, which is negligible. Thus output current is almost independent of input in this region and hence it is also called the **non-linear region.** Here BJT acts as an **off switch (open circuit).**

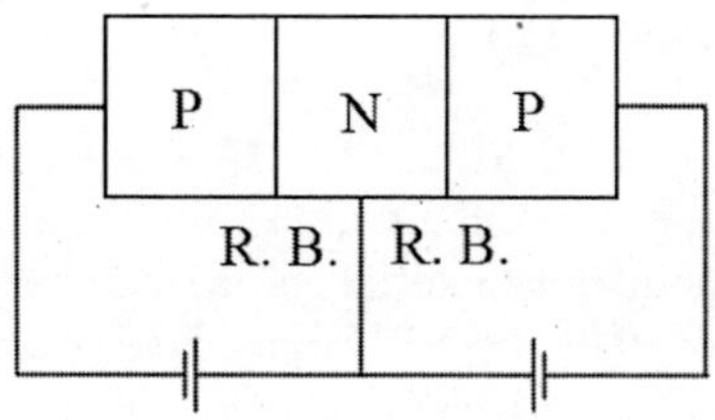

(b) **Active Region:** Here input junction is forward biased and the output junction is reversed biased. Here output current is directly proportional to the input current and so it is also called the linear region. It is also called the normal mode of operation. In this region, BJT can be used as an amplifier and as a constant current source.

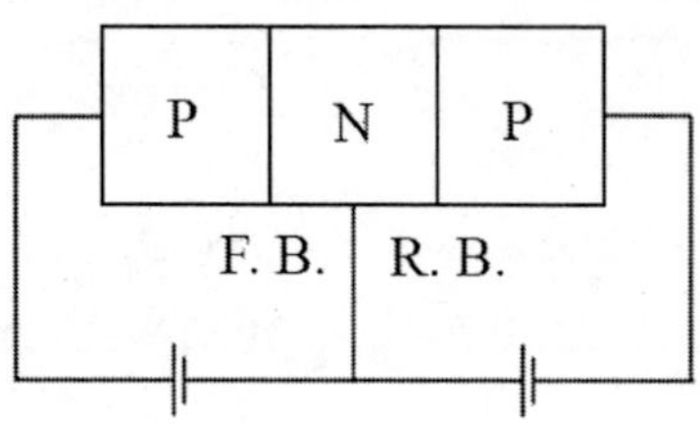

(c) **Saturation Region:** Here both the junctions are forward-biased. In this region output current is almost constant (almost independent of input) and hence it is also called a **non-linear region.** As both junctions are forward-biased, BJT almost acts as a short circuit and can be used as a **short switch.**

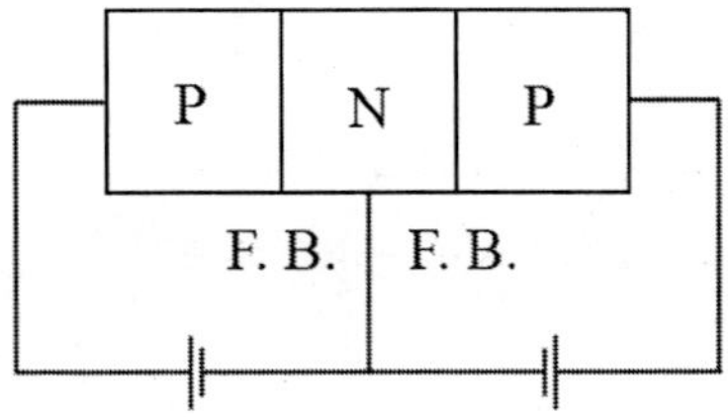

Depletion Region & Barrier Potential

The shaded portion shows a depletion region penetrating deeply into a lightly doped region. As the base is lightly doped compared to the emitter & collector, the depletion layer penetrates deeply into the base. Hence the distance between the emitter-base junction & collector-base junction is reduced. Barrier potential is positive on the ***n***- side and negative on the ***p***- side.

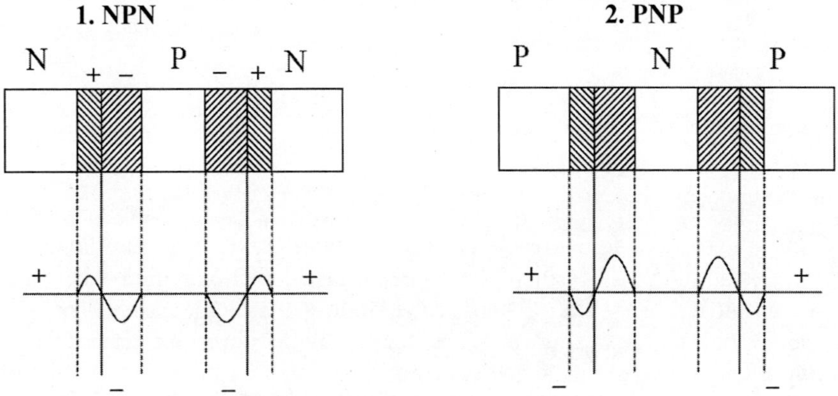

Fig. 2.2: Depletion Region and Barrier Potential

2.3 Transistor operation in linear mode

2.3.1 For NPN transistor

The working of BJT is studied in active regions. The emitter-base junction is forward-biased and the collector-base junction is reverse-biased. The forward bias at the emitter-base junction reduces barrier potential & causes electrons to flow from the emitter to the base. The electrons are emitted into the base, hence the name emitter. Holes flow from base to emitter. However, as the base is more lightly doped than the collector, almost all current flows across the emitter-base junction consisting of electrons. The reverse bias at the collector-base junction widens the depletion layer and penetrates deeply into the base.

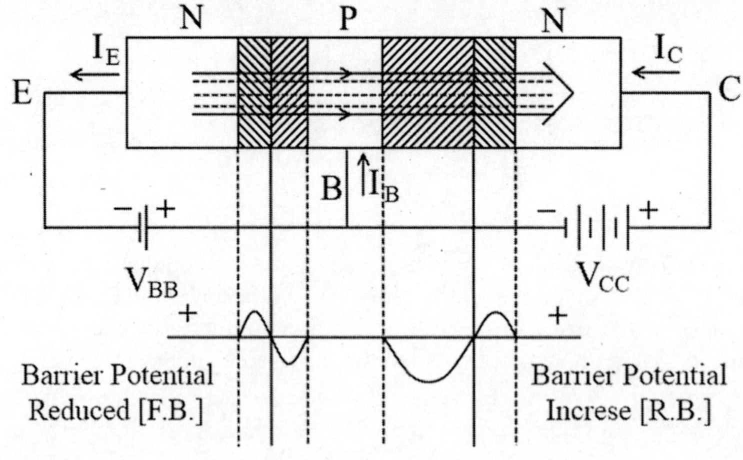

Fig. 2.3: Transistor Operation (NPN)

The electrons arrive close to a large positive electric field at the collector-base depletion region. Electrons are negatively charged and are drawn across the collector-base junction and they are collected in the collector.

Some of the charge carriers do not reach the collector but flow out via base connections around the base-emitter bias circuit. The path of the collector-base depletion region is very much shorter than the base terminal so only 2% of charge carriers flow out via the base. Also doping is light in the base so few holes in the base are available for recombination with electrons that are emitted by the emitter. The reserves bias at the collector-base junction opposes the flow of majority carriers and assists minority carrier flow. The majority of carriers are holes coming from the ***p***-side of a junction and electrons from the ***n***-side.

In the NPN configuration, electrons are charge carriers arriving at the collector-base junction, and the reverse is biased to assist its flow. If the forward bias of the emitter-base junction increases, barrier potential decreases, and more electrons flow to the base, hence current increases. If forward bias is reduced, barrier potential increases and current decreases. Hence a variation of the small forward bias voltage on the emitter-base junction controls the emitter and collector currents.

$$I_E = I_C + I_B$$

2.3.2 For PNP transistor

The operation is similar to the NPN configuration. Here the majority of charge carriers are holes. Holes are emitted from the ***p***-type emitter across the emitter-base junction into the base. The base is lightly doped and so fewer electrons are available to recombine with some holes to flow out via the base, however, most of them are drawn across the collector-base junction. The forward-biased emitter-base junction controls collector and emitter currents.

Although one type of charge carrier is in majority, two types of charge carriers are involved in current flows. Hence, these devices are called **Bipolar junction transistors**.

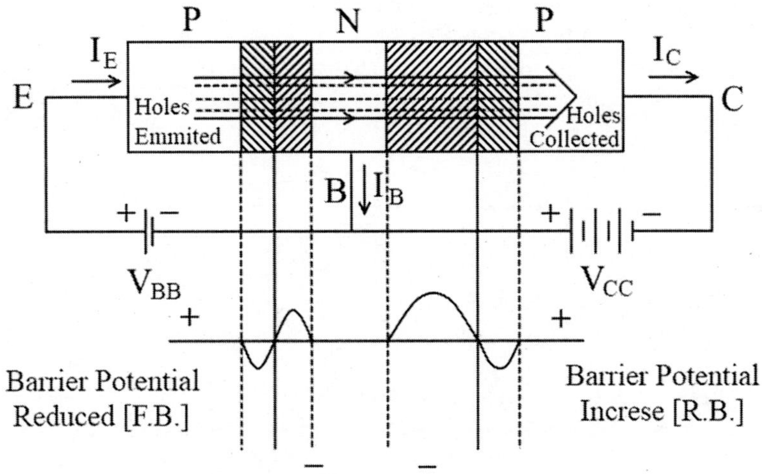

Fig. 2.4: Transistor Operation (PNP)

2.4. Transistor Configurations

A transistor can be used as a two-port network. Two port networks have two input and two output terminals. The BJT is used as a two-port network by grounding i.e. by making one of the terminals common between input and output.

Since any terminal can be grounded in BJT to use it as a two-port network it results in three types of configuration namely: (a) Common Base configuration, (b) Common Emitter configuration, and (c) Common Collector configuration.

2.4.1 Common Base Configuration (CB): Here base terminal is made common between input and output by grounding it as shown in the figure. V_{BB} is used to forward the bias base to the emitter junction while V_{CC} is used to reverse the bias base to the collector junction.

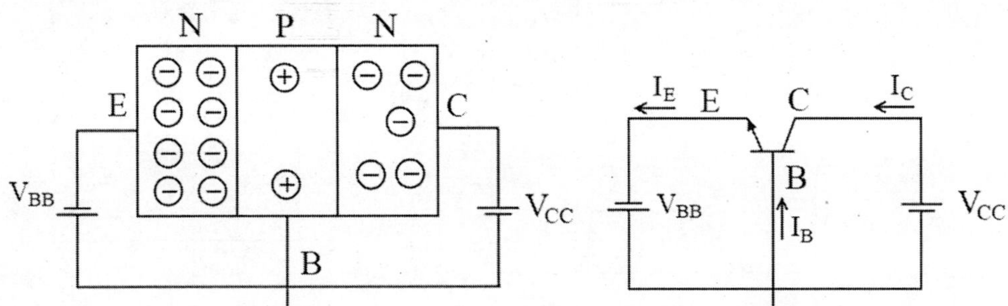

Fig. 2.5: Common Base Configuration

Forward biasing of the emitter junction results in the emission of charge carriers for the conduction process. When electrons travel from the emitter to the collector, free electrons recombine with holes available in the base region, while the remaining electrons are transferred to the collector.

$$I_C \propto I_E \quad \Rightarrow \quad I_C = \alpha\, I_E$$

$$\alpha = \frac{I_C}{I_E} = \frac{\text{Collector Current}}{\text{Emitter Current}}$$

Where α is proportionality constant and is called common base current amplification factor. Even if the emitter terminal is open-circuited, a small amount of current flows through the collector junction. This current is due to thermally generated charge carriers known as base-to-collector junction leakage current, I_{CBO}. This current continues even though the emitter junction is forward bias.

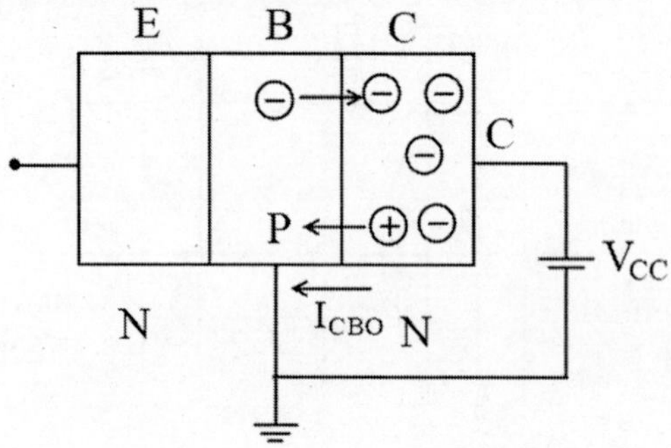

Fig. 2.6: I_{CBO} in CB Configuration

∴ The effective collector current is given by

$$I_C = \alpha I_E + I_{CBO}$$

2.4.2 Common Emitter Configuration (CE): In CE configuration emitter terminal is made common between input and output by grounding it. DC supply V_{BB} is used to forward bias base to emitter junction while V_{CC} reverse biases collector to base junction.

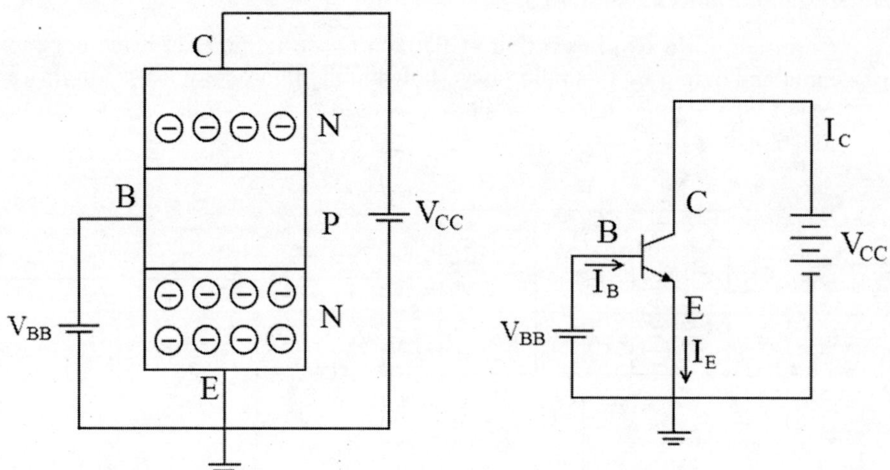

Fig. 2.7: Common Emitter Configuration

The supply voltage V_{BB} generates a few holes which in turn attract a large number of electrons from the emitter. These electrons arrive in the base region, few electrons recombine with holes while the remaining electrons are accelerated toward the collector by V_{CC}. Thus collector current is proportional to the base current i.e.

$$I_C \propto I_B$$

$$\therefore\ I_C = \beta I_B$$

where β is proportionality constant. It is called as common Emitter current gain factor.

Relation between α & β	Complete Current Equation
$I_E = I_C + I_B$ Divide by I_C $\frac{I_E}{I_C} = 1 + \frac{I_B}{I_C}$ $\frac{1}{\alpha} = 1 + \frac{1}{\beta}$ $\therefore \quad \alpha = \frac{\beta}{1+\beta}$ and $\beta = \frac{\alpha}{1-\alpha}$	$\because I_C = \alpha\, I_E + I_{CBO} \quad \& \quad I_E = I_C + I_B$ $\therefore \; I_C = \alpha(I_C + I_B) + I_{CBO}$ $I_C(1-\alpha) = \alpha\, I_B + I_{CBO}$ $I_C = \frac{\alpha}{1-\alpha} I_B + \frac{1}{1-\alpha} I_{CBO}$ $I_C = \beta\, I_B + (1+\beta) I_{CBO}$ Typically α = 0.98 to 0.998 and β = 100

2.4.3. Common Collector Configuration (CC): Here collector terminal is made common between input and output by grounding it. D.C. supply V_{BB} is used to forward bias the base to collector junction, while V_{EE} is used to reverse bias base to the emitter junction.

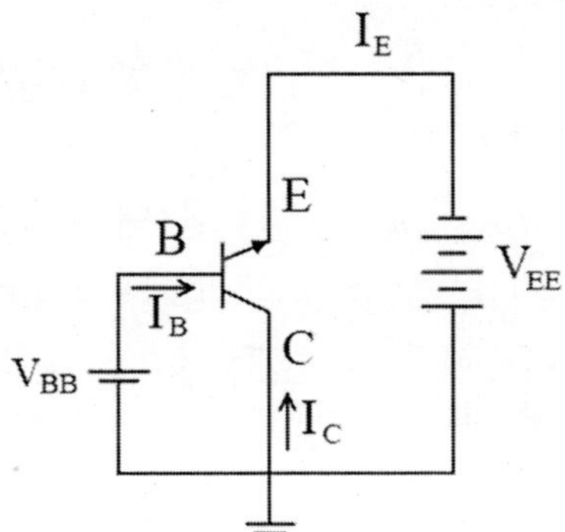

Fig. 2.8: Common Collector Configuration

Common collector current gain is-

$$\gamma = \frac{I_E}{I_B} = \frac{I_C + I_B}{I_B} = 1 + \beta$$

Practically above connections of supply are not used. Practically collector is grounded indirectly for AC signal and biasing is implemented only using a positive supply.

2.5 DC Load Line

The DC bias point or quiescent point (Q-Point) is the point on the DC load line which represents the current in a transistor and the voltage across it when no signal is applied i.e., it represents dc bias conditions. Biasing means the selection of DC or Q-point. The Q-point is selected based on the application. If a BJT is to be used as an amplifier or as a constant current source, then Q-point must be in the active region, while for switch or clipper applications the Q-point should be in the non-linear region. To understand the concept, let us consider the circuit shown in Fig. 2.9.

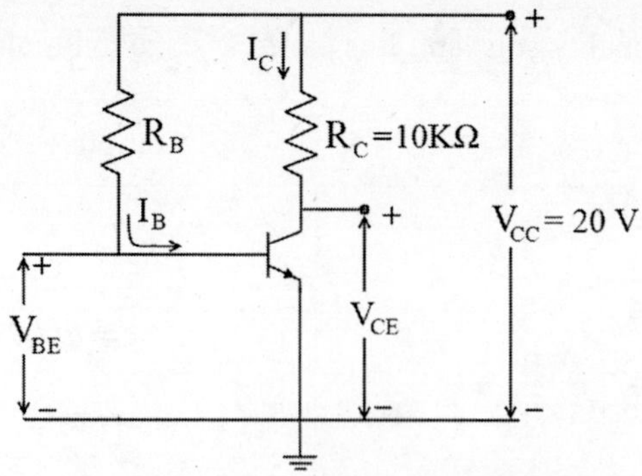

Fig. 2.9: Circuit under consideration

If this circuit is used as an amplifier, then input terminals are base and emitter and output terminals are collector and emitter i.e. CE configuration. The base-emitter junction is forward-biased and the collector-base junction is reverse-biased. From Fig. 2.9-

$$V_{CE} = V_{CC} - I_C R_C$$

If base voltage V_B is such that the transistor is not conducting i.e.

$$I_C = 0, \; V_{CE} = V_{CC} - [0 \times 10K] = 20V$$

When $\quad I_C = 0, \; V_{CE} = 20\,V$ [Let it be the point A]

if $\quad I_C = 2mA, \; V_{CE} = 0\,V$ [Let it be the point B]

Similarly, If $\quad I_C = 0.5mA, \; V_{CE} = 15\,V$ [Let it be the point C]

$I_C = 1mA, \; V_{CE} = 10\,V$ [Let it be the point D]

$I_C = 1.5mA, \; V_{CE} = 5\,V$ [Let it be the point E]

The line joining these points is straight and is the DC load line for R_C = 10KΩ. Fig. 2.10 depicts a typical DC load line for the above illustrations.

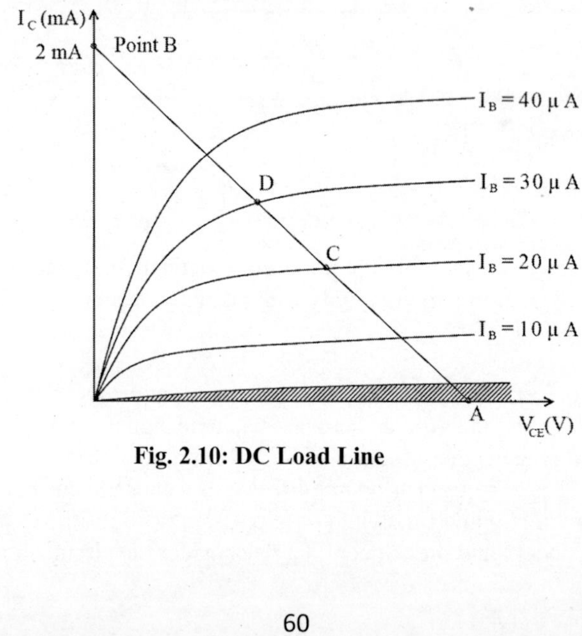

Fig. 2.10: DC Load Line

DC Load line is a straight line superimposed on the output characteristics of the transistor. It gives voltage across the transistor and current through the transistor. It can be plotted by drawing a straight line between points A and B only.

At point A,

let,

$I_C = 0$ mA and $V_{CE} = 20$ V and

At point B,

$V_{CE} = 0$ V and $I_C = 2$ mA

If a point is plotted such that $I_C = 1.5$ mA and $V_{CE} = 10$V, it will not appear on the load line. Hence, the load line shows that such a combination of voltage and current does not exist in the circuit. The load line shown in Fig. 2.10 is applicable for the conditions- $V_{CC} = 20$V and $R_C = 10$K. If either of these conditions is changed, the load line will change. Hence, the load line is sensitive to the variation in V_{cc} and R_c.

2.6 Transistor biasing

Biasing is a technique using which DC operating conditions are adjusted to operate the transistor in one of the three regions of output characteristics. Bipolar transistors must be properly biased to operate correctly. The bias circuit stabilizes the operating point of the transistor for variations in transistor characteristics and operating temperature. A bias network is selected to reduce the effects of device variability, temperature, and voltage changes. A bias circuit may be composed of only resistors or may include elements such as temperature-dependent resistors, diodes, or additional voltage sources, depending on the range of operating conditions expected.

The following are the common types of biasing circuits-

- Fixed bias
- Collector-to-base bias
- Emitter bias
- Voltage divider bias or potential divider

Potential divider bias:

The voltage divider is formed using external resistors R1 and R2. The voltage across R2 forward biases the emitter junction. By proper selection of resistors R1 and R2, the operating point of the transistor can be made independent of β. In this circuit, the voltage divider holds the base voltage fixed independent of the base current, provided the divider current is large compared to the base current. However, even with a fixed base voltage, collector current varies with temperature. Hence, an emitter resistor is added to stabilize the Q-point.

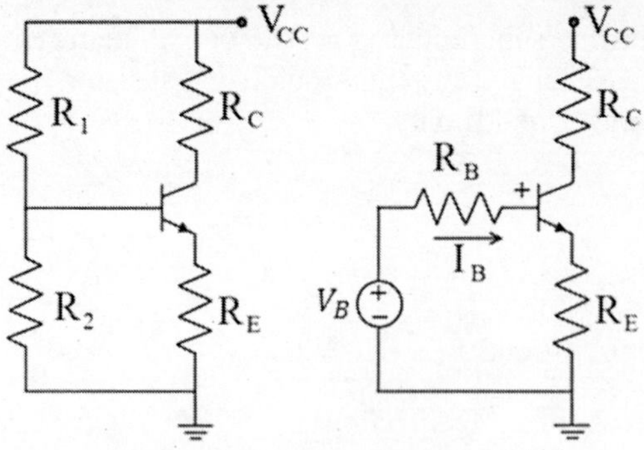

Fig. 2.11: Potential divider bias

The voltage divider configuration achieves the correct voltages through the use of resistors in certain patterns. In this circuit, the base voltage is given by-

$$V_B = \frac{V_{CC}R_2}{R_1 + R_2} \quad and \quad R_B = \frac{R_1 R_2}{R_1 + R_2}$$

Advantages:

- The operating point is almost independent of β variation.
- The operating point stabilized against a shift in temperature.

Disadvantages:

- In this circuit, to keep I_C independent of β the following condition must be met:

$$I_c = \frac{\frac{V_{CC}}{1 + {R_1}/{R_2}} - V_{be}}{R_e}$$

This is approximately the case if-

$$(\beta + 1)R_e \gg R_1 \parallel R_2$$

where $R_1 \parallel R_2$ denotes the equivalent resistance of R_1 and R_2 connected in parallel.

- As the β-value is fixed for a given transistor, this relation can be satisfied either by keeping R_E fairly large or making $R_1 \| R_2$ very low.
  - If R_E is of large value, a high V_{cc} is necessary. This increases cost as well as precautions necessary while handling.
  - If $R_1 \parallel R_2$ is low, either R_1 is low, or R_2 is low, or both are low. A low R_1 raises Vb closer to V_c, reducing the available swing in collector voltage, and limiting how large Rc can be made without driving the transistor out of active mode. A low R_2 lowers V_{be}, reducing the allowed collector current. Lowering both resistor values draws more current from the power supply and lowers the input resistance of the amplifier as seen from the base.

Usage:

The circuit's stability and merits as above make it widely used for linear circuits.

2.7. Thermal Stability

Transistor's characteristics are affected by temperature. The BE junction voltage V_{BE} and the collector to base saturation current I_{CBO} are very sensitive to temperature. As BE and CB are PN junctions the temperature effect is the same as the diode.

The temperature coefficient of V_{BE}, $\Delta V_{BE}/\Delta t$ is –1.8mv/°C for Si transistor and –2.02mv/°C for Ge. I_{CBO} doubles after every 10°C rise in temperature. As I_{CBO} increases, we know $I_C = \beta I_B + (1+\beta) I_{CBO}$, and thus I_C increases. It results in an increased power dissipation at the collector junction and a rise in the temperature of the CB junction. This generates more minority charge carriers, and I_{CBO} further increases. This effect is cumulative and I_C increases considerably. Hence, the DC operating point i.e. Q-point will shift. As I_C is increasing continuously, the CB junction overheats and burns out. This effect is called a Thermal runway.

Change in V_{BE} may produce a significant change in I_C and change Q-point. Because of the possibility of a thermal runaway, the changes in I_{CBO} are more important. The thermal stability of a circuit is assessed by deriving a stability factor.

(a) **Stability Factor (S):** It is defined as the ratio of the change in collector current to the change in the collector to base leakage current when V_{BE} and β are constant.

$$S = \frac{\Delta I_C}{\Delta I_{CO}} \quad \text{for constant } V_{BE} \text{ and } \beta$$

(b) **Thermal Stability factor (S'):** It is defined as the ratio of the change in collector current to the change in V_{BE} when collector to base leakage current and β are constant.

$$S' = \frac{\Delta I_C}{\Delta V_{BE}} \quad \text{for constant } I_{CO} \text{ and } \beta$$

(c) **Thermal Stability factor (S'')**

$$S'' = \frac{\Delta I_C}{\Delta \beta} \quad \text{for constant } I_{CO} \text{ and } V_{BE}$$

For better performance, the rate of change of I_C should be small concerning ΔI_{CO}, $\Delta\beta$ & ΔV_{BE}. So stability factor should be as small as possible. The larger the value of S, S', S", the more is the circuit thermally unstable.

2.7.1. Generalized Expression for S

We know that $I_C = \beta I_B + (1+\beta) I_{CBO}$, Differentiating with respect to I_C, we get

$$1 = \beta\frac{\partial I_B}{\partial I_C} + (1+\beta)\frac{\partial I_{CBO}}{\partial I_C}$$

$$S = \frac{1+\beta}{1-\beta\frac{\partial I_B}{\partial I_C}}$$

To evaluate S for any circuit an expression for $\frac{\partial I_B}{\partial I_C}$ must be derived.

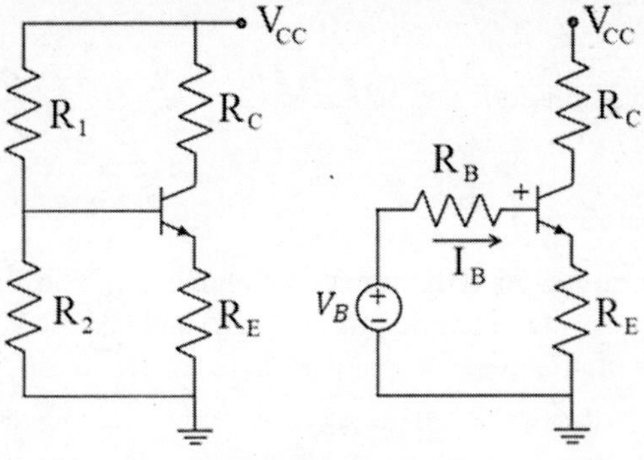

Fig. 2.12: Self bias circuit

$$V_B = \frac{V_{CC}R_2}{R_1 + R_2} \quad and \quad R_B = \frac{R_1R_2}{R_1 + R_2}$$

KVL to input loop

$$V_B = I_BR_B + V_{BE} + (I_C + I_B)R_E$$

Differentiate w. r. t. I_C

$$0 = (R_B + R_E)\frac{\partial I_B}{\partial I_C} + R_E + 0$$

$$\Rightarrow \frac{\partial I_B}{\partial I_C} = \frac{-R_E}{R_B + R_E}$$

$$S = \frac{1+\beta}{1 + \frac{\beta}{1+K}} \quad \text{where} \quad K = \frac{R_B}{R_E}$$

For better stability, K = R_B/R_E should be as small as possible. i.e. R_B should be small and R_E should be large. For example, S varies from 7 to 7.7. Among all biasing circuits, this circuit gives the highest thermal stability and hence is widely used in practical circuits.

2.7.2. Bias Compensation

Generally, the stabilization and compensation techniques are used to provide maximum bias and thermal stabilization to the transistors. Diodes, thermistors, and sensistors can be used to compensate for variations in current.

2.7.2.1. Thermistor Compensation: Thermistor R_T having a negative temperature coefficient is connected in parallel with R_2. An increase in temperature will decrease the resistance of the thermistor will decrease V_{BE}, reducing I_B and I_C. Thus compensation is achieved.

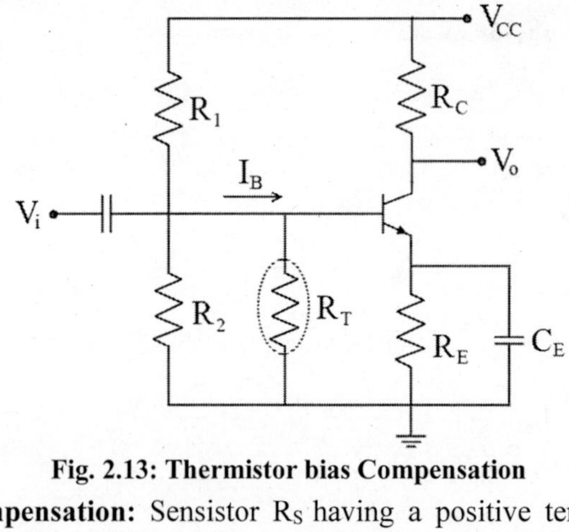

Fig. 2.13: Thermistor bias Compensation

2.7.2.2. Sensistor Compensation: Sensistor R_S having a positive temperature coefficient is connected in parallel with R_1. An increase in temperature will increases the resistance of the sensistor will decrease V_{BE}, reducing I_B and I_C. Thus compensation is achieved.

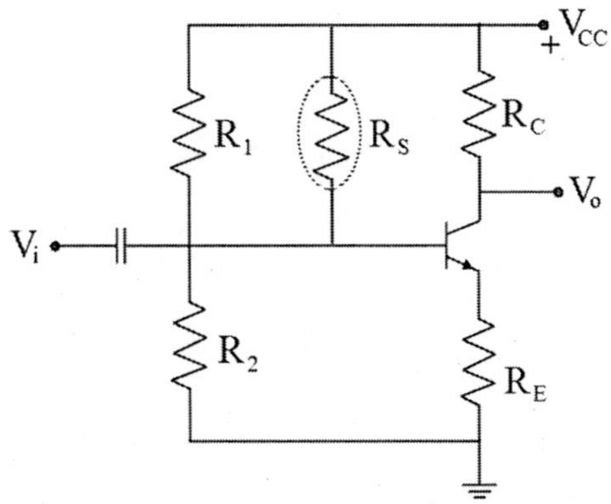

Fig. 2.14: Sensistor bias Compensation

2.8. Transistor as an Amplifier

A transistor acts as an amplifier by raising the strength of a weak signal. The DC bias voltage applied to the emitter-base junction makes it remain in a forward-biased condition. This forward bias is maintained regardless of the polarity of the signal. Fig. 2.15 shows a transistor configuration when connected as an amplifier.

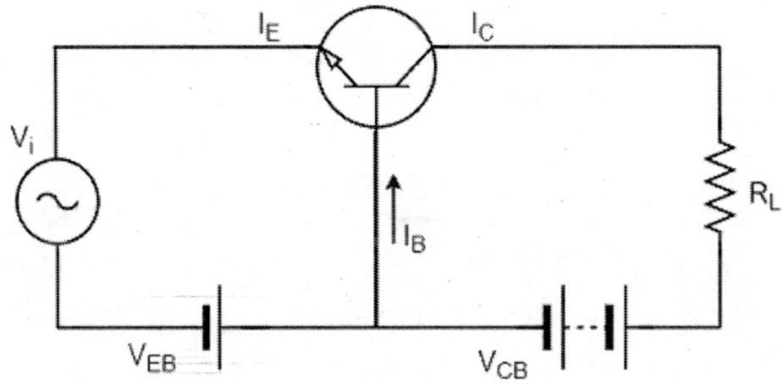

Fig: 2.15: Transistor as an Amplifier

The low resistance in the input circuit lets any small change in the input signal result in an appreciable change in the output. The emitter current caused by the input signal contributes to the collector current, which then flows through the load resistor R_L, resulting in a large voltage drop across it. Thus a small input voltage results in a large output voltage, which shows that the transistor works as an amplifier.

Let there be a change of 0.1v in the input voltage being applied, which further produces a change of 1mA in the emitter current. This emitter current will produce a change in collector current, which would also be 1mA.

A load resistance of 5kΩ placed in the collector would produce a voltage of -

$$5\ k\Omega \times 1\ mA = 5V$$

Hence, it is observed that a change of 0.1v in the input gives a change of 5V in the output, which means the voltage level of the signal is amplified.

2.9. The Field Effect Transistor (FET)

A FET is a three-terminal unipolar semiconductor device that has very similar characteristics to the Bipolar Transistor. The three terminals are namely Gate, Source, and Drain. The FETs are constructed with no PN-junctions within the main current carrying path between the Drain and the Source terminals (i.e. similar in function to the collector and the emitter respectively of the BJT). The current path between these two terminals is called the *channel*. A typical channel may be made of either a P-type or an N-type semiconductor material. The control of the current flowing in this channel is achieved by varying the voltage applied to the Gate.

BJTs are *bipolar* devices as they can operate with both types of charge carriers i.e. holes and electrons. The FETs can operate with only one type of charge carrier i.e. either holes(P-channel) or electrons (N-channel). Hence, they are also known as *unipolar* devices.

The FETs use the voltage applied to the input terminal, known as the Gate terminal, to control the current flowing through them. It results in an output current that is proportional to the input voltage. As the FETs depend on an electric field generated by the input Gate voltage, it is a voltage-controlled device. The input impedance of FETs is very high compared to the BJTs. The very high input impedance makes them very sensitive to input voltage signals.

The FETs are classified as the Junction Field Effect Transistor(JFET) and the Insulated-gate Field Effect Transistor (IGFET). The IGFETs are also known as Metal Oxide Semiconductor Field Effect Transistors (MOSFET).

2.9.1. Junction Field effect transistors(JFET)

It consists of an ***n***-type or ***p***-type silicon bar which is lightly doped. Two terminals are taken out from the two sides of the silicon bar, known as source 'S' and drain 'D'. The source and drain are identical and they can be interchanged. On the other two sides of the silicon bar, two *p*-type or *n*-type regions are heavily doped. These two regions are internally connected and form a third terminal of JFET known as Gate 'G'. Due to these *p*-regions, two p-n junctions are formed (Drain-Gate, Source-Gate).

The region between two p-n junctions is called as the channel which allows the flow of the current. If the silicon bar is made of *n*-type material, it is called as n-channel JFET. If the silicon bar is made of *p*-type material, then it is called as *p*-channel JFET.

The channel of the N-channel JFETs is doped with donor impurities. Hence, the flow of current through this channel is due to the flow of electrons. Similarly, the channel of the P-channel JFET is doped with acceptor impurities. Hence, the flow of current through this channel is due to the flow of holes. Due to the higher mobility of electrons as compared to the holes, an N-channel JFET has a better channel conductivity as compared to the p-channel JFETs. Fig. 2.16 shows the construction and symbol of an *n*-channel and *p*-channel JFET.

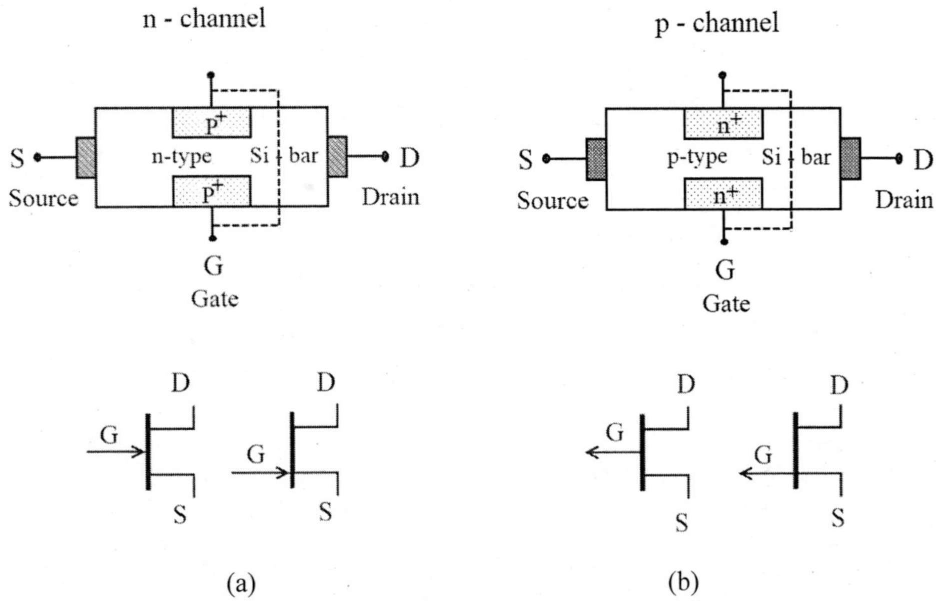

Fig. 2.16: Construction and symbol of an (a) n-channel and (b) p-channel JFET

2.9.2. Operation of a JFET (n-channel)

In Fig. 2.17, a typical ***n***-channel JFET is shown with applied voltage polarities. The supply V_{DD} provides drain-source voltage V_{DS}. Due to this applied voltage, Drain current I_D flows from drain to source. Here, one needs to understand the behavior of drain current with gate to source and drain to source voltage.

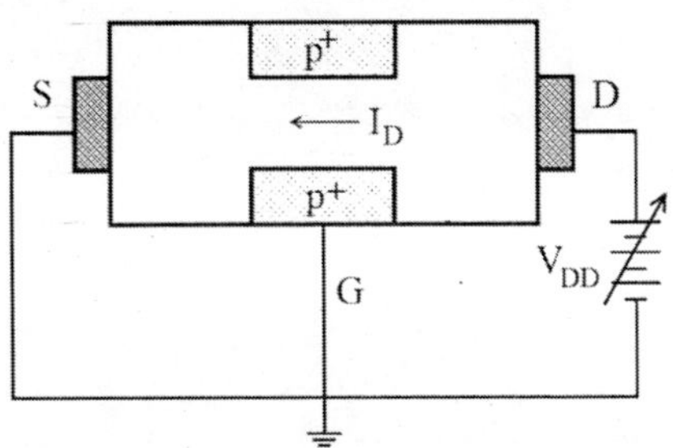

Fig. 2.17: *N* - Channel FET with applied drain voltage polarities.

JFET operation can be explained by different values of V_{GS} as-

When $V_{GS} = 0V$.

When $V_{GS} = 0V$ and $V_{DS} = Possitive,$ the two p-n junctions (D-G, S-G) are reverse biased. As V_{DS} is positive, the electrons from the channel and near the two p-n junctions are attracted to the drain terminal. Hence drain current I_D flows through the channel from drain to source (conventional direction).

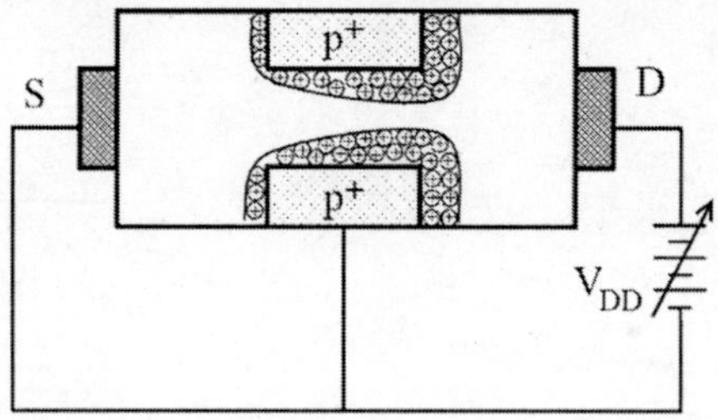

Fig. 2.18: The rise in depletion width with applied $V_{DS}(V_{DD})$

Due to the motion of electrons to the drain, the channel region (n-type) gets positively biased and the gate is zero-biased. Thus $P^{+}N$ junctions are reverse biased and hence depletion region is created near the two p-n junctions as shown in Fig. 2.18.

As any semiconductor material(n type or p type) is resistive, I_D causes a voltage drop across the channel. In the portion of the channel between the gate and the source, I_D causes a voltage drop which biases the Gate concerning that part of the channel close to the gate. Hence, point A is positive concerning the source. i.e. Source is negative concerning point A. Since the gate is connected to the source, the gate region is negative concerning point A by a voltage V_A.

Thus depletion region penetrates the channel at point A by an amount proportional to V_A. Between point B and the source voltage drop along the channel is V_B, which is less than V_A. Therefore, at point B, the gate is at $-V_B$ concerning channel, and depletion region penetrates less than point A. Similarly $V_C < V_B$ and penetration at point C is less than at point A and point B. Hence penetration of the depletion region is maximum on the drain side than on source side as shown in Fig 2.19.

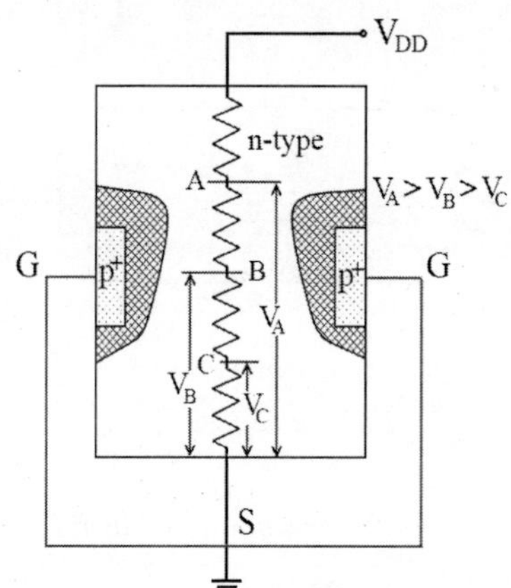

Fig. 2.19: Depletion Region development due to internal voltage drop

As V_{DS} is further increased depletion region increases and channel size decreases and a stage is reached at which the channel width becomes very small and the drain current I_D remains constant because a fixed number of electrons are attracted to the drain. It is called as a pinch-off region.

This constant current is called as a drain to source current (I_{DSS}) when the gate and source are shorted ($V_{GS} = 0$). I_{DSS} is also called drain to source saturation current. The characteristics curve for $V_{GS} = 0$ is shown below.

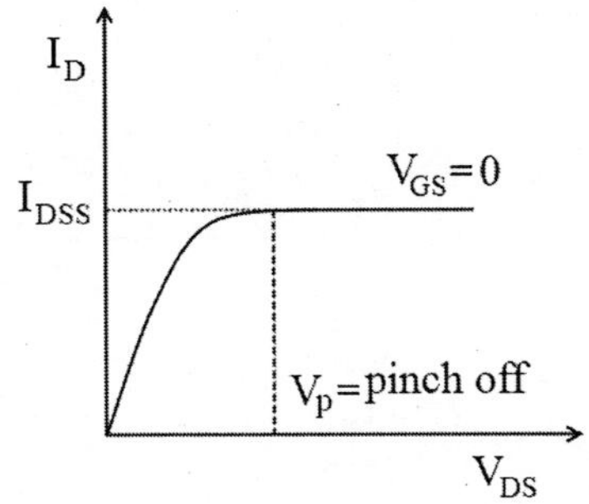

When $V_{GS} = -ve\ (-1V)$

When negative gate-source voltage is applied it reverses biases in both p-n junctions. Hence small depletion region is formed in the channel due to negative V_{GS}. As V_{DS} is increased in the positive direction more electrons are attracted to the drain terminal. Hence depletion region also increases, and the drain current I_D increases. At one stage I_D becomes constant since the channel width becomes very small through which constant numbers of electrons are attracted to the drain terminal. In this case, the constant level of I_D is below that at $V_{GS} = 0$. Since depletion region is formed initially due to the applied negative gate to source voltage and the remaining depletion region is formed due to the motion of electrons to the drain terminal. The characteristics are shown below

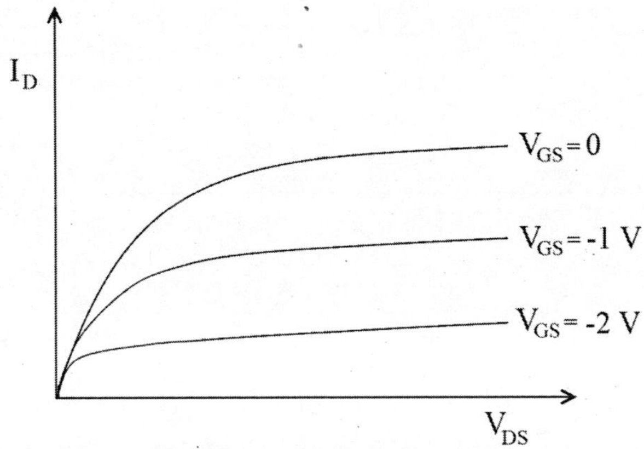

When $V_{GS} = -ve$ (more negative or $V_{GS} = V_P$)

When negative V_{GS} is increased the depletion region also increases in size and the drain current I_D decreases. At one stage depletion region is completely formed across the channel and I_D becomes zero in this case the depletion region is only due to negative V_{GS} which completely blocks the channel. Hence electrons cannot move from source to drain. Hence I_D becomes zero. It is called the pinch-off region as shown in Fig.2.20.

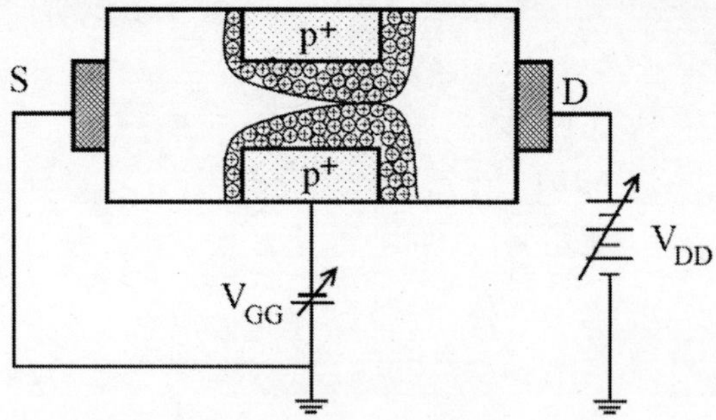

Fig. 2.20: Pinch-off occurs at $V_{GS} = V_P$

Pinch-off voltage: The negative gate to source voltage at which the depletion region is completely formed across the channel and hence drain current I_D reduces to zero, is called pinch-off voltage. Fig. 2.21 shows the output characteristics of a JFET.

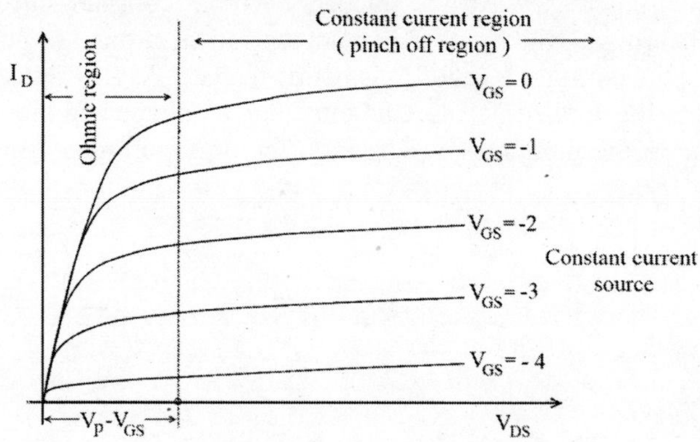

Fig. 2.21: Output Characteristics

2.9.3. FET Parameters

a) Trans-conductance (g_m): It is the slope of the transfer characteristics

$$g_m = \left.\frac{\Delta I_D}{\Delta V_{GS}}\right|_{V_{DS} = \text{constant}}$$

It is called as AC gain of FET. It is measured in Siemens (s) or mhos (℧).

$$\because \; I_D = I_{DSS}\left(1 - \frac{V_{GS}}{V_P}\right)^2$$

Differentiate with respect to V_{GS}

$$\frac{\partial I_D}{\partial V_{GS}} = -\frac{2\, I_{DSS}}{V_P}\left[1 - \frac{V_{GS}}{V_P}\right]$$

$$\therefore \; g_m = g_m\left[1 - \frac{V_{GS}}{V_P}\right]$$

Where $g_m = -\frac{2I_{DSS}}{V_P}$

g_{mo} → Maximum transconductance

b) Drain Source (ON) resistance (R_{DS}) or $R_{D\,(ON)}$

R_{DS} is the DC resistance of the channel when the depletion region is absent and when the device is biased on in the ohmic region of characteristics.

$$I_D \times R_{DS(\text{on})} = V_{DS(\text{on})}$$

This is similar to the V_{CE} (sat) of BJT. FET can function as a Voltage Variable Resistor (VVR) in the ohmic region of characteristics. Here resistance can be varied by varying V_{GS}

c) Drain resistance: In the constant current region the reciprocal of the slope of the characteristics is the drain resistance rd.

$$r_d = \left.\frac{\Delta V_{DS}}{\Delta I_D}\right| V_{GS} = \text{Constant}$$

As V_{DS} increases by a large value the change in I_D is very small (ideally zero). Hence the drain resistance rd is very high (ideally infinite)

***d)* Amplification Factor (μ):** It is the product of the transconductance and drain resistance of the device.

$$\mu = \left.\frac{\Delta V_{DS}}{\Delta V_{GS}}\right| I_D = Constant$$

$$\mu = g_m \times r_d$$

e) **Gate cut-off current (I_{GSS}):** The gate channel junction in a FET is a PN junction, it is normally reverse biased, and current flows because of minority carriers. It is also called a Gate-Source cut-off current I_{GSS} or gate reverse current.

f) ***Input Resistance: (R_{GS})*** It is the resistance of the reverse biased gate channel junction and is inversely proportional to I_{GSS}. It is very high.

2.9.4. Comparison between BJT and FET

Parameter	**BJT**	**JFET**
Polarity	Bipolar	Unipolar
Carriers	Holes and electrons	Either holes or electrons
Nature of operation control	Current controlled	Voltage Controlled
Input Impedance	Very low	Very High
Gain Bandwidth product	High	Low
Thermal Stability	Low	High
Gain	High	Low

2.10. MOSFET (Metal oxide semiconductor field effect transistor)

It is also called as an insulated gate field effect transistor (IGFET). There are two types of MOSFETs-

1. Depletion MOSFET (D- MOSFET)
2. Enhancement MOSFET (E- MOSFET)

2.10.1. D- MOSFET (*n*-channel)

Construction:

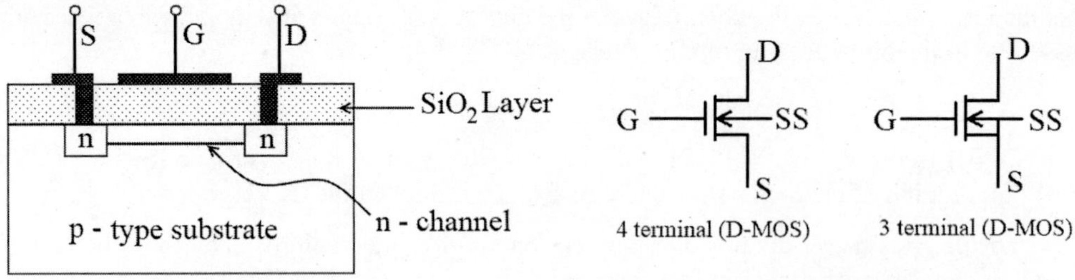

Construction *Symbol*

Fig. 2.22: Construction and symbol of a D-MOSFET

There is a p-type substrate. On the p-type substrate, two n-type regions are heavily doped, which forms the source and drain terminals of the MOSFET. Source and drain terminals are internally connected by lightly doped n-channel.

On the p-type substrate, Silicon dioxide (SiO_2) layer is placed which acts as a dielectric or insulator. Aluminum contacts are taken out for the source and drain. Aluminum metal is deposited over the SiO_2 layer, which forms the gate. Hence we can say that aluminum is deposited on oxide (SiO_2) which is deposited over a semiconductor (P-type). Hence the name metal oxide semiconductor field effect Transistor.

Operation:

When $V_{GS} = 0$

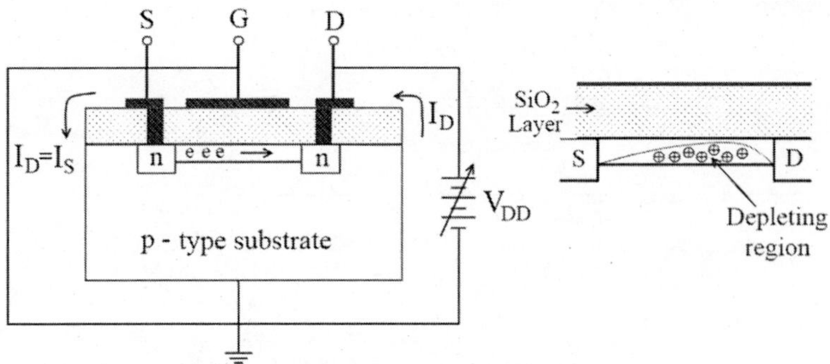

Fig.2.23: Operation of a D-MOSFET at V_{GS}=0

Here the gate is insulated from conducting channel by (SiO_2) layer. Hence it is called as an insulated gate field effect Transistor. The operation of n-channel D-MOSFET is similar to n-channel JFET when $V_{GS} = 0$. The drain terminal is connected to the positive terminal of the battery. Gate and source are shorted. The substrate terminal is connected to the ground. As the drain terminal is connected to the positive potential, electrons from the channel are attracted toward the drain. Hence, positive donor ions in the channel are uncovered and a small depletion region is formed in the channel. The electrons travel to the drain terminal and then to the positive terminal of the battery. From the negative terminal, they enter in the source terminal. Hence, drain current flows from the drain to the source (conventional direction).

As V_{DD} is increased, more electrons are attracted to the drain terminal, hence depletion region also increases. At one balancing stage, the channel width becomes very small through which fixed numbers of electrons are attracted to the drain, hence I_D remains constant. It is denoted by I_{DSS} i.e. drain to source current when gate and source are shorted.

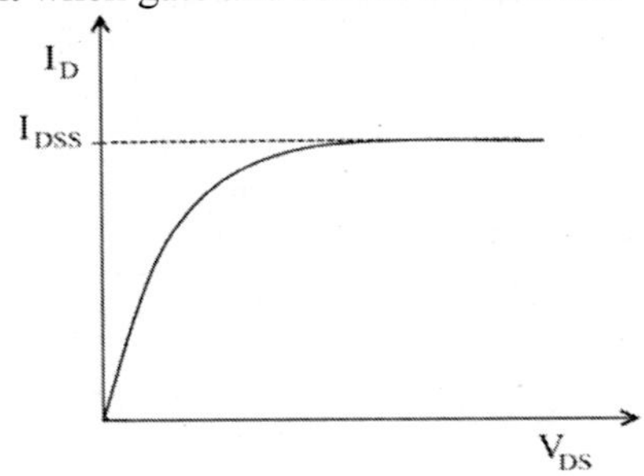

V_{GS} = negative

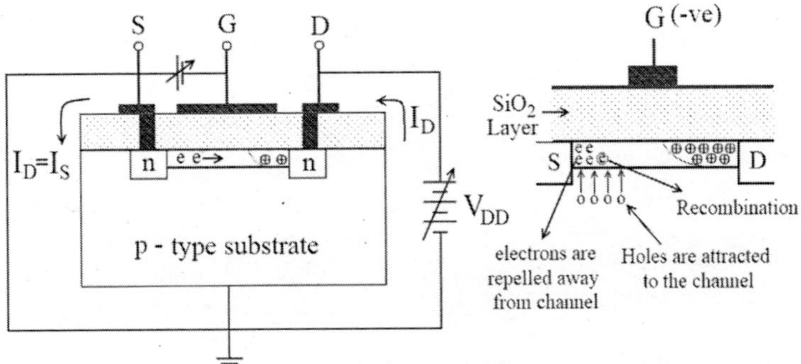

Fig. 2.24: Operation of a D-MOSFET at V_{GS}= negative

When V_{GS} is negative, it attracts positive charges in the channel and repels the electrons from the channel. Due to the recombination of the charge carriers, the number of electrons available for conduction decreases. A small depletion region is formed in the channel due to negative V_{GS}.

As V_{GS} becomes more negative, the depletion region in the channel increases. Hence, I_D decreases. At the point of operation under this condition, the channel is completely blocked, and the drain current I_D becomes zero. This is called as a pinch-off condition of the channel. The negative V_{GS} at which the channel is completely blocked and the drain current I_D becomes zero is called pinch-off voltage (V_P).

V_{GS} = possitive

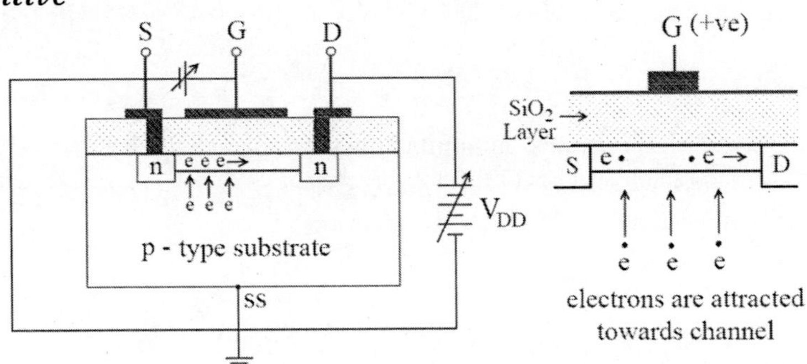

Fig. 2.25: Operation of a D-MOSFET at V_{GS}= positive

When V_{GS} is positive, the negative charges are attracted to the channel. The negative charges attracted in the channel are minority carriers. Hence, the number of electrons in the channel increases.

When positive V_{GS} is increased, these electrons along with the electrons from the channel are attracted to the drain terminal. As a result, the drain current I_D increases. In turn, the depletion region also increases. At one point of operation under this condition, the channel width becomes very narrow and I_D remains constant such that $I_D > I_{DSS}$.

When V_{GS} is positive, the channel is never completely blocked as the depletion region is entirely due to the motion of electrons to the drain. Hence with positive V_{GS}, conduction is enhanced and D-MOSFET can work as E-MOSFET.

Drain source characteristic:

It is a plot of drain current I_D versus positive drain-source voltage for a fixed value of V_{GS}. As V_{GS} becomes more and more negative, I_D goes on decreasing. However, as V_{GS} is made more positive, drain current I_D goes on increasing. The drain-source characteristics and transfer characteristics are shown in Fig. 2.36.

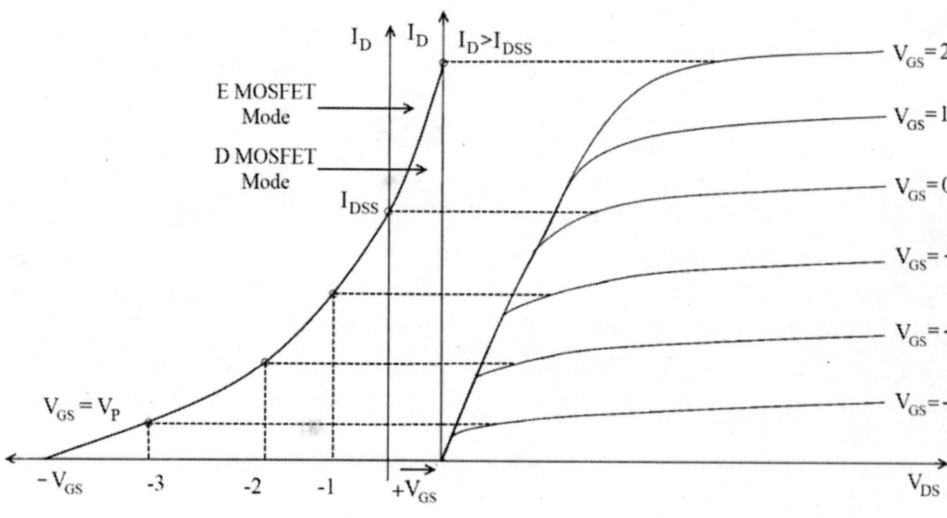

Fig. 2.26: The drain-source characteristics and transfer characteristics

p-channel D-MOSFET :

The p-channel D-MOSFET operates on similar lines to an n-channel D-MOSFET. Fig. 2.37 shows the symbol of a p-channel D-MOSFET.

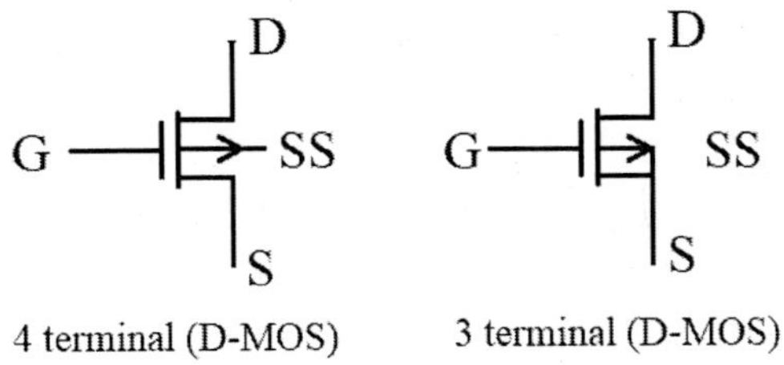

Fig. 2.27: Symbol fr a *p*-channel D-MOSFET

2.10.2. E-MOSFET (*n*-channel)

Construction:

It has a lightly doped *p*-type substrate. On the p-type substrate, two *n*-regions are heavily doped. There is no conducting channel between the source and drain when the device is constructed. Channel is enhanced or formed with appropriate gate-source voltage. Hence, it is known as the enhancement or E-MOSFET. Fig.2.38 shows the construction and symbol of an E-MOSFET.

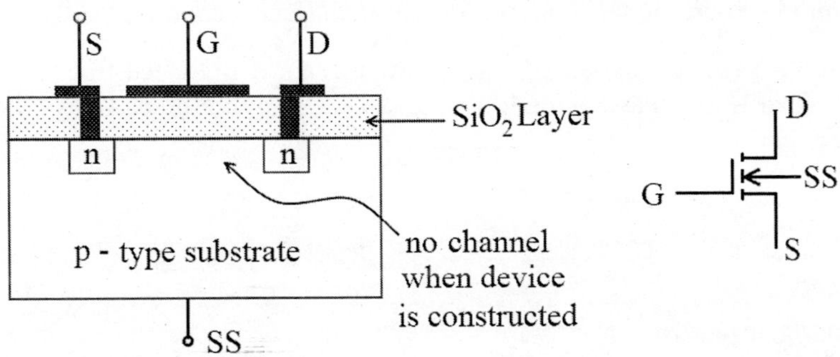

Fig. 2.28: Construction and symbol of an E-MOSFET

Operation :

When $V_{GS} = 0$

When $V_{GS} = 0$ and the drain terminal is connected to the positive terminal of a battery, the drain current $I_D = 0$ as there is no channel between the source and drain. Hence, electrons from the source cannot travel to the drain terminal.

V_{GS} = negative,

When V_{GS} is negative, due to the capacitive effect, positive charges will be attracted to the SiO2 layer. The drain current I_D is zero because there is no conducting channel (n-type) between the n-type source and the drain. Therefore *E-MOSFET* cannot operate for $V_{GS} = 0V$ and $V_{GS} - ve$.

V_{GS} = positive,

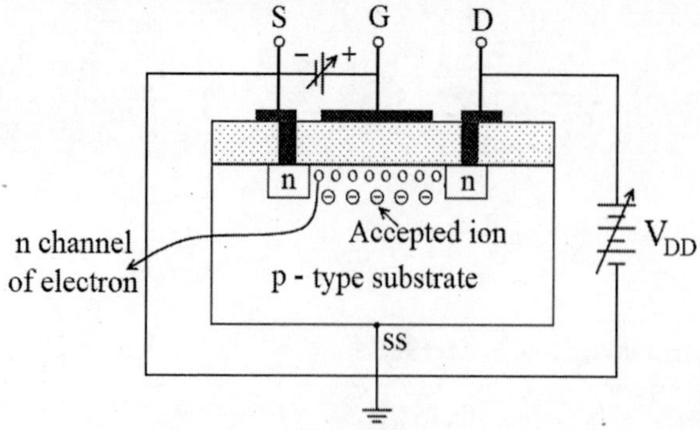

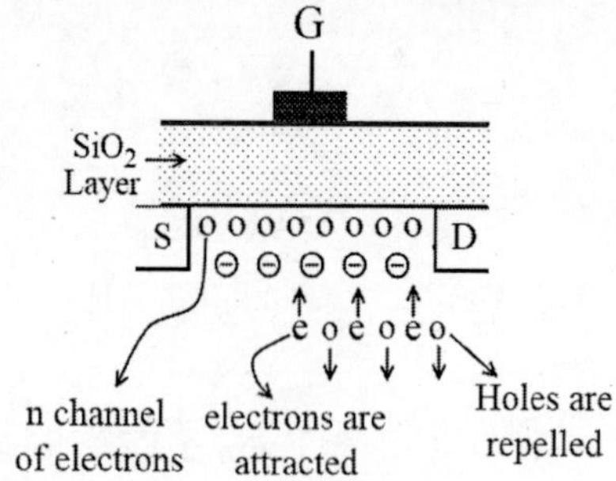

When V_{GS} is made positive, the negative charges are attracted to the SiO2 layer and repel the positive charges away from it. Hence, holes from a p-type substrate are repelled away from the SiO2 layer and minority carriers(electrons) from a substrate are attracted to the SiO2 layer. The negative acceptor ions in the *p*-type substrate are uncovered because holes move away from the SiO2 layer. The minority carriers(electrons) will cover these negative ions due to positive V_{GS}. After one stage, a layer of electrons is formed from the source to drain near the SiO2 layer. Thus n-channel is created or channel is enhanced. The electrons travel from the source to the drain via this channel. Hence, drain current I_D flows.

The positive gate-source voltage at which the *n*-channel is created from the source to the drain terminal is called as a threshold voltage V_T. Up to threshold voltage, $I_D = 0$. The *n*-channel of electrons is also called as an inversion layer.

If V_{DS} is made more positive, the constant level of I_D goes on increasing along with increasing V_{GS}.

Drain Source Characteristics:

The drain-source graph is a graph of $I_D V_S$, V_{DS} when V_{GS} = constant. As V_{GS} is made more *+ve*, I_D increases.

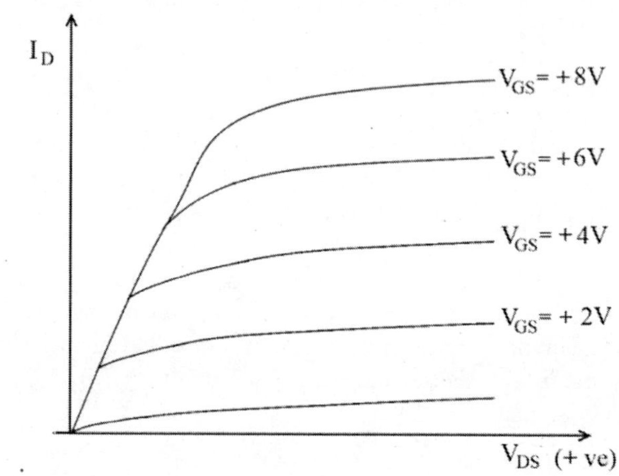

Fig. 2.29: Drain Source Characteristics

2.11 CMOS

The term CMOS stands for 'Complementary Metal Oxide Semiconductor'. This is one of the most popular technologies in the computer chip design industry and it is broadly used today to form integrated circuits in numerous and varied applications. Today's computer memories, CPUs, and cell phones make use of this technology due to several key advantages. This technology makes use of both *p*-channel and *n*-channel semiconductor devices. One of the most popular MOSFET technologies available today is the Complementary MOS or CMOS technology. This is the dominant semiconductor technology for microprocessors, microcontroller chips, memories like RAM, ROM, EEPROM, and application-specific integrated circuits (ASICs).

The main advantage of CMOS over NMOS and BIPOLAR technology is the much smaller power dissipation. Unlike NMOS or BIPOLAR circuits, a Complementary MOS circuit has almost no static power dissipation. Power is only dissipated in case the circuit switches. This allows the integration of more CMOS gates on an IC than in NMOS or bipolar technology, resulting in much better performance. Complementary Metal Oxide Semiconductor transistor consists of P-channel MOS (PMOS) and N-channel MOS (NMOS). Fig.2.40 shows a basic CMOS configuration.

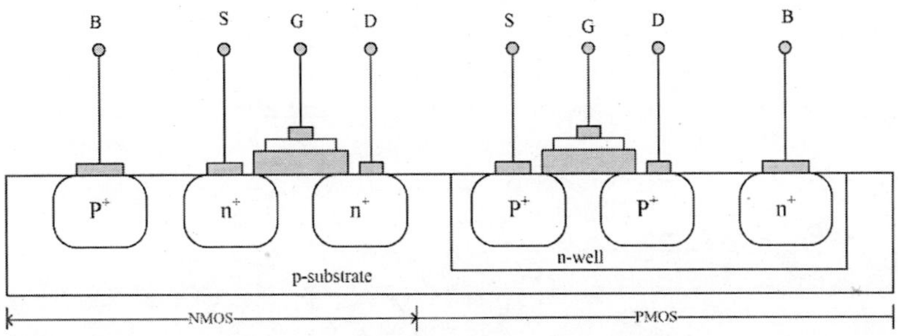

Fig. 2.30: Basic CMOS configuration

NMOS

NMOS is built on a p-type substrate with an n-type source and drain diffused on it. In NMOS, the majority of carriers are electrons. When a high voltage is applied to the gate, the NMOS will conduct. Similarly, when a low voltage is applied to the gate, NMOS will not conduct. NMOS is considered to be faster than PMOS, since the carriers in NMOS, which are electrons, travel twice as fast as the holes.

PMOS

p-channel MOSFET consists of P-type Source and Drain diffused on an N-type substrate. The majority of carriers are holes. When a high voltage is applied to the gate, the PMOS will not conduct. When a low voltage is applied to the gate, the PMOS will conduct. PMOS devices are more immune to noise than NMOS devices.

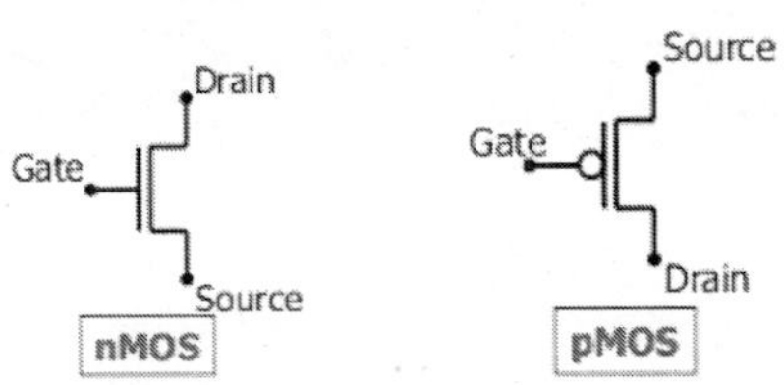

Fig. 2.31: Symbol of an NMOS and PMOS transistor

CMOS Working Principle:

In CMOS technology, both N-type and P-type transistors are used to design logic functions. The same signal which turns ON a transistor of one type is used to turn OFF a transistor of the other type. This characteristic allows the design of logic devices using only simple switches, without the need for a pull-up resistor.

In CMOS logic gates a collection of n-type MOSFETs is arranged in a pull-down network between the output and the low-voltage power supply rail (Vss or quite often ground). Instead of the load resistor of NMOS logic gates, CMOS logic gates have a collection of p-type MOSFETs in a pull-up network between the output and the higher-voltage rail (often named Vdd).

Thus, if both a p-type and n-type transistor has their gates connected to the same input, the p-type MOSFET will be ON when the n-type MOSFET is OFF, and vice-versa. The networks are arranged such that one is ON and the other OFF for any input pattern as shown in Fig. 2.32.

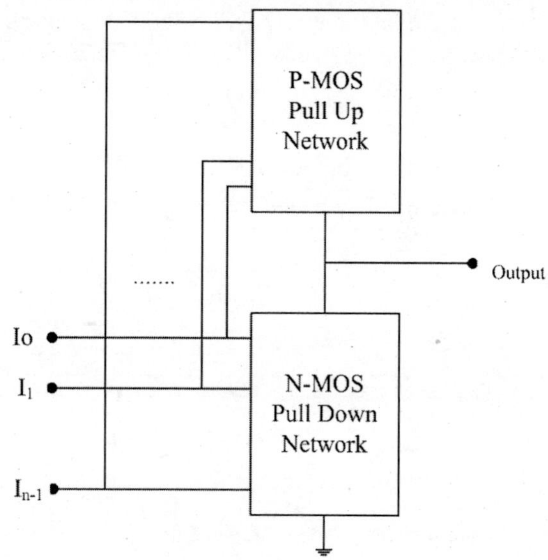

Fig.2.32: CMOS using Pull Up and Pull Down

CMOS offers relatively high speed, low power dissipation, and high noise margins in both states, and will operate over a wide range of source and input voltages (provided the source voltage is fixed).

CMOS Inverter:

The inverter circuit is shown in Fig. 2.33. It consists of PMOS and NMOS FET. Input A serves as the gate voltage for both transistors. The NMOS transistor has input from Vss (ground) and the PMOS transistor has input from Vdd. Terminal Y is the output. When a high voltage (~ Vdd) is given at the input terminal (A) of the inverter, the PMOS becomes an open circuit, and NMOS is switched OFF so the output will be pulled down to Vss.

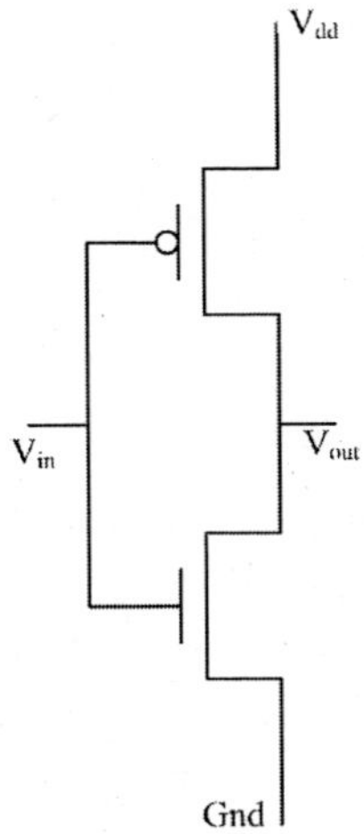

Fig. 2.33: CMOS Inverter

When a low-level voltage (<Vdd, ~0v) is applied to the inverter, the NMOS switched OFF and PMOS switched ON. Hence, the output becomes Vdd or the circuit is pulled up to Vdd.

INPUT	LOGIC INPUT	OUTPUT	LOGIC OUTPUT
0 v	0	Vdd	1
Vdd	1	0 v	0

Advantages :

- The main benefits of CMOS over TTL are good noise margin as well as less power consumption.
- It uses a single power supply like + VDD
- Input impedance is high
- CMOS logic uses less power whenever it is held in a set state
- Power dissipation is negligible
- Fan out is high
- TTL compatibility
- Stability of temperature
- Noise immunity is good
- Compact in size

Disadvantages:

- The cost will be increased once the processing steps increases, however, it can be resolved.
- The packing density of CMOS is low as compared with NMOS.
- MOS chips should be secured from getting static charges by placing the leads shorted otherwise; the static charges obtained within leads will damage the chip. This problem can be solved by including protective circuits otherwise devices.
- Another drawback of the CMOS inverter is that it utilizes two transistors as opposed to one NMOS to build an inverter, which means that the CMOS uses more space over the chip as compared with the NMOS.

====

Solved Examples

1. Determine the dc bias voltage V_{CE} and the current I_C for the voltage-divider configuration of Fig. 2.34.

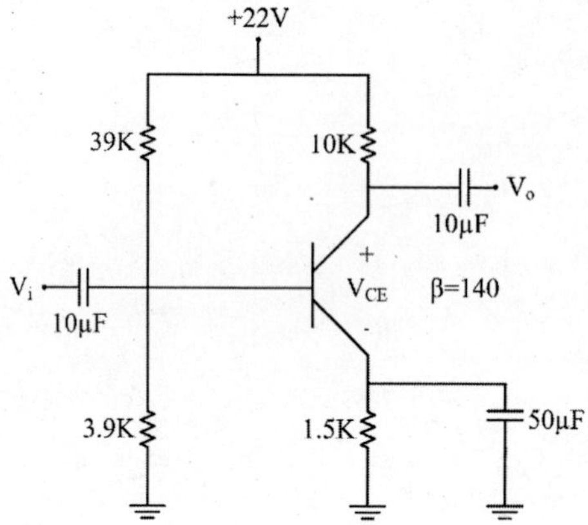

Fig. 2.34: Circuit diagram

Sol:

$$R_{TH} = R_1 \parallel R_2$$
$$= \frac{(39\ K\Omega)(3.9\ K\Omega)}{39\ K\Omega + 3.9\ K\Omega} = 3.55\ K\Omega$$

$$E_{TH} = \frac{R_2 V_{CC}}{R_1 + R_2}$$
$$= \frac{(3.9\ K\Omega)(22\ V)}{39\ K\Omega + 3.9\ K\Omega} = 2\ V$$

$$I_B = \frac{E_{TH} - V_{BE}}{R_{TH} + (\beta + 1)R_E}$$

$$= \frac{2\ V - 0.7\ V}{3.55\ K\Omega + (141)(1.5\ K\Omega)} = \frac{1.3\ V}{3.55\ K\Omega + 211.5\ K\Omega} = 6.05\ \mu A$$

$$I_C = \beta I_B$$
$$= (140)(6.05\ \mu A)$$
$$= 0.85\ mA$$

$$V_{CE} = V_{CC} - I_C(R_C + R_E)$$
$$= 22V - (0.85mA)(10K\Omega + 1.5K\Omega)$$
$$= 22V - 9.78V = 12.22\ V\ \ [\boldsymbol{ANS}]$$

2. For potential divider bias circuit, estimate variation in Q-point when $V_{CC} = 12V$, $R_C = 1.2K\Omega$, $R_E = 1 \times 10^3\Omega$, $R_2 = 10 \times 10^3\Omega$, $R_1 = 42 \times 10^3\ \Omega$, $\beta = 100, 50, 150$, $V_{BE} = 0.7V$ and also find the variation in Q point.

Sol.: Circuit Diagram

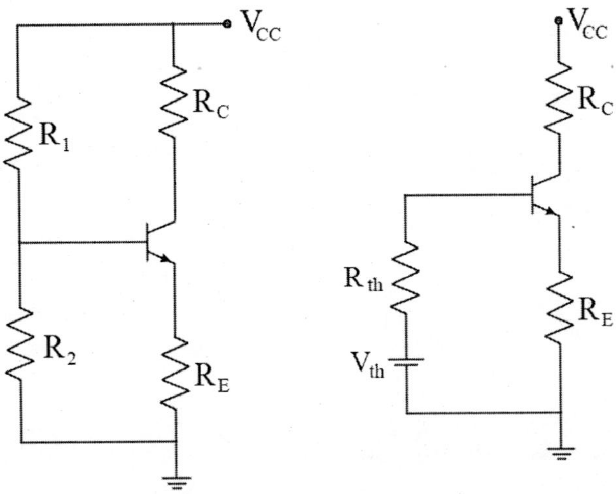

$$R_{TH} = \frac{R_1 R_2}{R_1 + R_2}$$

$$R_{TH} = \frac{42 \times 10^3 \times 10 \times 10^3}{42 \times 10^3 + 10 \times 10^3} = 8.07 \times 10^3\Omega$$

$$V_{TH} = \frac{V_{CC} R_2}{R_1 + R_2} = \frac{12 \times 10 \times 10^3}{42 \times 10^3 + 10 \times 10^3} = 2.30V$$

For β=100

$$I_B = \frac{V_{TH} - V_{BE}}{R_{TH} + (1 + \beta)R_E} = \frac{2.30 - 0.7}{8.07 \times 10^3 + (101) \times 1 \times 10^3} = 14.66 \times 10^{-6}A$$

$$I_C = \beta I_B = 100 \times 14.66 \times 10^{-6} = 1.466 \times 10^{-3}A$$

$$V_{CE} = V_{CC} - I_C(R_C + R_E) = 12 - 1.466 \times 10^{-3} \times (1.2 \times 10^3 + 1 \times 10^3)$$

$$V_{CE} = 8.772V$$

For $\beta_1 = 50$

$$I_{B1} = \frac{2.30 - 0.7}{8.07 \times 10^3 + 51 \times 1 \times 10^3} = 27.08 \times 10^{-6}A$$

$$I_{C1} = \beta_1 I_{B1} = 50 \times 27.08 \times 10^{-6} = 1.35 \times 10^{-3}A$$

$$V_{CE1} = V_{CC} - I_{C1}(R_C + R_E)$$

$$V_{CE1} = 12 - 1.35 \times 10^{-3}(1 \times 10^3 + 1.2 \times 10^3 = 9.020V$$

For $\beta_2 = 150$

$$I_{B2} = \frac{2.30 - 0.7}{8.07 \times 10^3 + 151 \times 1 \times 10^3} = 10.05 \times 10^{-6} A$$

$$I_{C2} = \beta_2 I_{B2} = 150 \times 10.05 \times 10^{-6} = 1.50 \times 10^{-3} A$$

$$V_{CE2} = V_{CC} - I_{C2}(R_C + R_E) = 8.680V$$

$$\% I_C = \frac{I_{C2} - I_{C1}}{I_C} \times 100 = 10.23\%$$

$$\% V_{CE} = \frac{V_{CE_1} - V_{CE_2}}{V_{CE}} \times 100 = 3.86\%$$

3. For the biasing circuit shown below, determine the Q-point and locate it on output characteristics

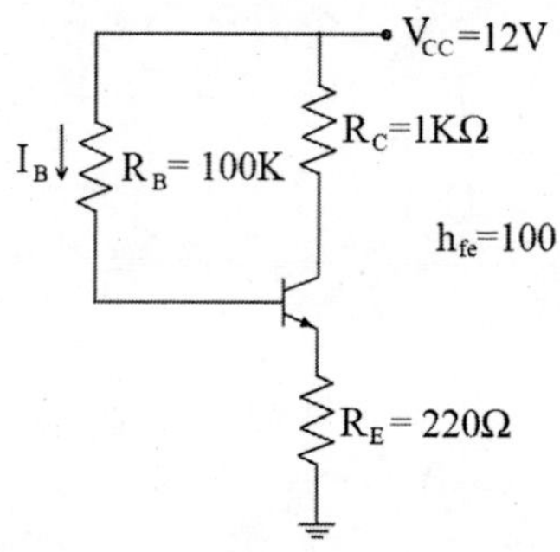

Sol.:

Let $V_{CE}(std) = 0.2V$ and $V_{BE} = 0.7V$

a) Appling KVL to the output-

$$V_{CC} = I_C R_C + V_{CE} + I_E R_E \quad \&$$

$$I_C \leq I_E, \; I_C = \frac{V_{CC} - V_{CE}}{R_C + R_E}$$

$$I_{C(sat)} = \frac{V_{CC} - V_{CE(sat)}}{R_C + R_E} = 9.67mA$$

From the input loop:

$$I_B = \frac{V_{CC} - V_{BE}}{R_B + (1 + \beta)R_E} = 0.0924\, mA \quad \&$$

$$I_C = \beta I_B = 9.24\, mA = I_{CQ}$$

To find the region of operation of the transistor-

If $I_{Bmin} \geq \frac{I_{Csat}}{\beta} E$ → BJT is in saturation otherwise is inactive

Here β $I_{Bmin} < I_{C(sat)}$

Hence, the transistor is in the active region.

DC load line

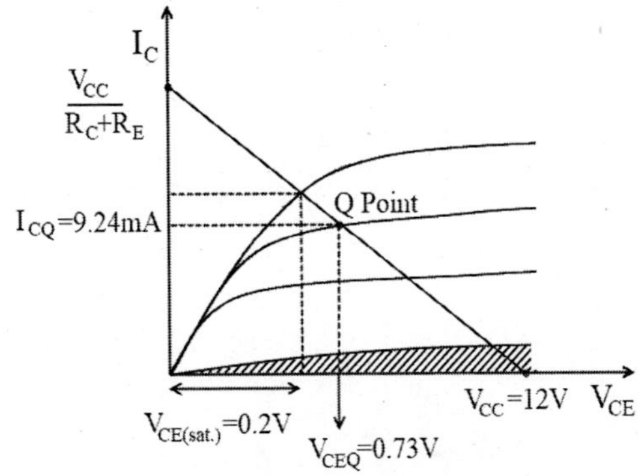

Exercise Questions

1. What is a bipolar junction transistor? How are its terminals named?
2. Discuss the amplifying action of a BJT.
3. What is the major difference between a bipolar and a unipolar device?
4. How must the two transistor junctions be biased for proper transistor amplifier operation?
5. What is the source of the leakage current in a transistor?
6. Derive the relationship between α and β.
7. Calculate the values of I_C and I_E for a transistor with $\alpha_{DC} = 0.9$ and $I_{CBO} = 5\mu A$. I_B is measured as $19.90\ \mu A$.

   [Ans.: $I_C = 2.49\ mA,\ I_E = 2.5\ mA$]
8. Calculate the emitter current for $\alpha_{DC} = 0.99$ and $I_{CBO} = 20\mu A$.
9. What is meant by Q point?
10. Discuss the importance of a DC load line in transistor operation.
11. What is biasing of a transistor? State its need in the operation of a transistor.
12. List the different biasing techniques? Describe the potential divider biasing scheme in detail.
13. What is a thermal runaway? How can be avoided?
14. Define stability factor. Derive the expression for the stability factor.
15. A self-biased silicon CE transistor amplifier operates with $V_{CC} = 10V, R_1 = 10K\Omega, R_2 = 6\Omega, R_C = 2K\Omega, R_E = 3K\Omega$ and $\beta = 100$, find (i) the coordinates of the operating point, and (ii) the stability factor.
16. Repeat Q. 11 for a germanium transistor.
17. Determine the operating point for the circuit of a potential divider bias arrangement with $R_2 = R_C = 10K\Omega, R_E = 2K\Omega$ and $R_1 = 80K\Omega$.
18. Compare BJT and FET.
19. Describe the construction and working of a JFET

20. List the classification of a MOSFET. Describe each type in detail.
21. What is the significant difference between the construction of an enhancement-type MOSFET and a depletion-type MOSFET?

Self Study Questions

1. What names are applied to the two types of BJT transistors? Sketch the basic construction of each and label the various minority and majority carriers in each. Draw the graphic symbol next to each. Is any of this information altered by changing from a silicon to a germanium base?
2. Define α, β and γ of a transistor. Show how they are related to each other.
3. Discuss the rationale of the stability factor. What would it seem more reasonable to call this an instability factor?
4. Write a comparative note on the amplifying action of a BJT and FET.
5. Compare the different configurations of a BJT.
6. Which of the transistor currents is always the largest? Which is always the smallest? Which two currents are relatively close in magnitude?
7. In what ways is the construction of a depletion-type MOSFET similar to that of a JFET? In what ways is it different?
8. Explain in your own words why the application of a positive voltage to the gate of an n-channel depletion-type MOSFET will result in a drain current exceeding I_{DSS}.
9. Research CMOS logic at your local or college library, and describe the range of applications and basic advantages of the approach.

MCQs

1. In a PNP transistor, the current carriers are ....
   a) acceptor ions
   b) donor ions
   c) free electrons
   d) holes
2. The collector of a transistor is .... doped
   a) heavily
   b) moderately
   c) lightly
   d) none of the above
3. A transistor is a .... operated device
   a) current
   b) voltage
   c) both voltage and current
   d) none of the above
4. The emitter of a transistor is .... doped
   a) lightly

b) heavily
c) moderately
d) none of the above

5. At the base-emitter junctions of a transistor, one finds ....
   a) a reverse bias
   b) a wide depletion layer
   c) low resistance
   d) none of the above
6. Most of the majority carriers are from the emitter ....
   a) recombine in the base
   b) recombine in the emitter
   c) pass through the base region to the collector
   d) none of the above
7. The output impedance of a transistor is ....
   a) high
   b) zero
   c) low
   d) very low
8. The most commonly used transistor arrangement is .... arrangement
   a) common emitter
   b) common base
   c) common collector
   d) none of the above
9. As the temperature of a transistor goes up, the base-emitter resistance ....
   a) decreases
   b) increases
   c) remains the same
   d) none of the above
10. In a transistor, the signal is transferred from a .... circuit
    a) high resistance to low resistance
    b) low resistance to high resistance
    c) high resistance to high resistance
    d) low resistance to low resistance
11. The arrow in the symbol of a transistor indicates the direction of ......
    a) electron current in the emitter
    b) electron current in the collector
    c) hole current in the emitter
    d) donor ion current
12. The most commonly used semiconductor in the manufacture of a transistor is ....
    a) germanium
    b) silicon
    c) carbon

d) none of the above

13. In a transistor, the collector current is controlled by ....
    a) collector voltage
    b) base current
    c) collector resistance
    d) all of the above

14. A JFET has three terminals, namely ....
    a) cathode, anode, grid
    b) emitter, base, collector
    c) source, gate, drain
    d) none of the above

15. A JFET is also called .... transistor
    a) unipolar
    b) bipolar
    c) unijunction
    d) none of the above

16. A JFET is a .... driven device
    a) current
    b) voltage
    c) both current and voltage
    d) none of the above

17. The gate of a JFET is .... biased
    a) reverse
    b) forward
    c) reverse as well as forward
    d) none of the above

18. The input impedance of a JFET is .... that of an ordinary transistor
    a) equal to
    b) less than
    c) more than
    d) none of the above

19. If the reverse bias on the gate of a JFET is increased, then the width of the conducting channel ....
    a) is decreased
    b) is increased
    c) remains the same
    d) none of the above

20. A MOSFET has .... terminals
    a) two
    b) five
    c) four
    d) three

21. A MOSFET can be operated with ....
    a) negative gate voltage only

b) positive gate voltage only
c) positive as well as a negative gate voltage
d) none of the above

22. A JFET has .... power gain
a) small
b) very high
c) very small
d) none of the above

23. The input control parameter of a JFET is ....
a) gate voltage
b) source voltage
c) drain voltage
d) gate current

24. A JFET has high input impedance because ....
a) it is made of semiconductor material
b) input is reverse biased
c) of impurity atoms
d) none of the above

25. The two important advantages of a JFET are ....
a) high input impedance and square-law property
b) inexpensive and high output impedance
c) low input impedance and high output impedance
d) none of the above

26. A MOSFET is sometimes called .... JFET
a) many gate
b) open gate
c) insulated gate
d) shorted gate

27. Which of the following devices has the highest input impedance?
a) JFET
b) MOSFET
c) Crystal diode
d) ordinary transistor

28. The pinch-off voltage of a JFET is about ....
a) 5 V
b) 0.6 V
c) 15 V
d) 25 V

29. The gate voltage in a JFET at which drain current becomes zero is called .... voltage
a) saturation
b) pinch-off
c) active
d) cut-off

30. The full form of CMOS is ___________
a) Capacitive metal oxide semiconductor
b) Capacitive metallic oxide semiconductor
c) Complementary metal oxide semiconductor
d) Complemented metal oxide semiconductor

31. CMOS technology is used in ___________
a) Inverter
b) Microprocessor
c) Digital logic
d) Both microprocessor and digital logic

32. Two important characteristics of CMOS devices are ___________
a) High noise immunity
b) Low static power consumption
c) High resistivity
d) Both high noise immunity and low static power consumption

33. CMOS behaves as a/an.......
a) Adder
b) Subtractor
c) Inverter
d) Comparator

34. An important characteristic of a CMOS circuit is the ___________
a) Noise immunity
b) Duality
c) Symmetricity
d) Noise Margin

35. CMOS logic dissipates _______ power than NMOS logic circuits.
a) More
b) Less
c) Equal
d) Very High

Answer Keys:

1.	**d**	11.	**c**	21.	**c**	31.	**d**
2.	**b**	12.	**d**	22.	**b**	32.	**d**
3.	**a**	13.	**b**	23.	**d**	33.	**c**
4.	**b**	14.	**c**	24.	**b**	34.	**b**
5.	**c**	15.	**a**	25.	**a**	35.	**b**
6.	**c**	16.	**b**	26.	**c**		
7.	**a**	17.	**a**	27.	**b**		
8.	**a**	18.	**c**	28.	**a**		
9.	**a**	19.	**a**	29.	**b**		
10.	**b**	20.	**d**	30.	**c**		

3. Feedback Amplifier and Oscillators

RATIONALE

The concept of feedback was originally introduced in 1934 by H. S. Black, an Electronics Engineer, for building an amplifier with a gain that is insensitive to changes in the amplifier parameters. Since then, this notion has played an important role in many areas of Engineering, especially in the design and development of feedback amplifiers and oscillators. In this chapter, we will study the details of feedback amplifiers and oscillators.

UNIT OUTCOMES

U1-O1: Unit-1 Learning Outcome-1

To know about the concept of feedback and its role in device operations

U1-O2: Unit-1 Learning Outcome-2

To know about the classification and operational characteristics of different types of negative feedback amplifiers

U1-O3: Unit-1 Learning Outcome-3

To study the classification, construction, and operation of different types of oscillators

LEARNING OBJECTIVES

LO1: To understand the concept of feedback in amplifiers.

LO2: To study the feedback connection types

LO3: To study the input and output impedance of feedback amplifiers

LO4: To study the general characteristics of a negative feedback amplifier

LO5: To study different types of Oscillators

MAPPING THE UNIT OUTCOMES WITH THE COURSE OUTCOMES

Unit Outcome	EXPECTED MAPPING WITH COURSE OUTCOMES (1- Weak Correlation; 2- Medium correlation; 3- Strong Correlation)						
	CO-1	CO-2	CO-3	CO-4	CO-5	CO-6	CO-7
U3-O1	1	--	--	3	--	--	--
U3-O2	1	--	--	3	--	--	--
U3-O3	1	--	--	3	--	--	--

Interesting Facts:

1. The concept of feedback was originally introduced in 1934 by H. S. Black, an Electronics Engineer, for building an amplifier with a gain that is insensitive to changes in the amplifier parameters. Since then, this notion has played an important role in many areas of Engineering.
2. Paul Voigt patented a negative feedback amplifier in January 1924, though his theory lacked detail.
3. Harold Stephen Black independently invented the negative-feedback amplifier while he was a passenger on the Lackawanna Ferry on his way to work at Bell Laboratories on August 2, 1927 (US patent 2,102,671, issued in 1937).
4. The first practical oscillators were based on electric arcs, which were used for lighting in the 19th century. The current through an arc light is unstable due to its negative resistance and often breaks into spontaneous oscillations, causing the arc to make hissing, humming, or howling sounds which had been noticed by Humphry Davy in 1821, Benjamin Silliman in 1822, Auguste Arthur de la Rive in 1846, and David Edward Hughes in 1878. Ernst Lecher in 1888 showed that the current through an electric arc could be oscillatory.
5. An oscillator was built by Elihu Thomson in 1892 by placing an LC-tuned circuit in parallel with an electric arc and included a magnetic blowout.
6. Mathematical conditions for feedback oscillations, now called the Barkhausen criterion, were derived by Heinrich Georg Barkhausen in 1921.
7. The first analysis of a nonlinear electronic oscillator model, the *Van der Pol oscillator*, was done by *Balthasar van der Pol* in 1927. He showed that the stability of the oscillations (limit cycles) in actual oscillators was due to the nonlinearity of the amplifying device.

Video Resources:

Sr	Title	URL	QR Code
1.	Concept of feedback in the amplifiers	https://www.youtube.com/watch?v=Qu0-QXvpZrA	

2.	How does Oscillator work?	https://www.youtube.com/watch?v=XVS8Puf4tiw	

3.1. The concept of feedback

Feedback is the process of combining the output of a system with its input. In negative feedback systems, the system's output is subtracted from its input, and in the case of a positive feedback system; the output is added to its input. Negative feedback stabilizes the system whereas positive feedback can make the system unstable. The feedback mechanism plays a vital role in the working of an amplifier. An amplifier is known as a feedback amplifier when its output influences the input. It is also known as the closed-loop amplifier as the path through the 'A' and 'β' networks form a loop. Here, 'A' networks are the amplifiers and a 'β' network consists of feedback elements. Figure 3.1 shows the block diagram of a single-loop feedback system.

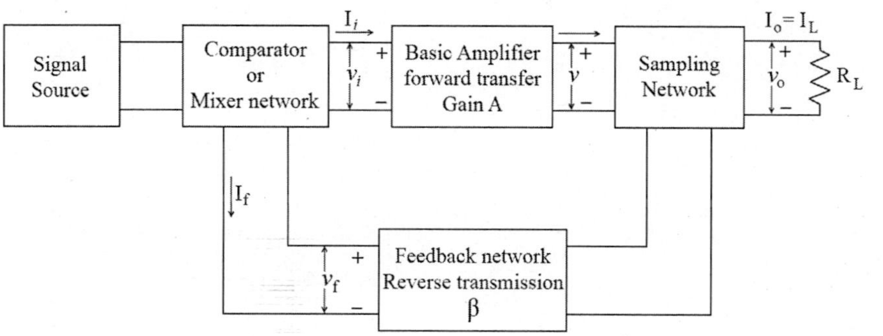

Fig 3.1. A single-loop feedback system

Signal Source: This is either a signal voltage V_S in series with a resister R_S, or a signal current I_S in parallel with a resistor R_S.

Comparator or Mixer Network: Two types of mixer or comparator networks exists-

1. Series mixing (figure 3.2 a): The network in which the input voltage is compared with feedback voltage by connecting the feedback network in series with the input is known as a series mixer.
2. Shunt mixing (figure 3.2 b): The network in which the input current is compared with the feedback current by connecting the feedback network in the shunt across input is known as a shunt mixer.

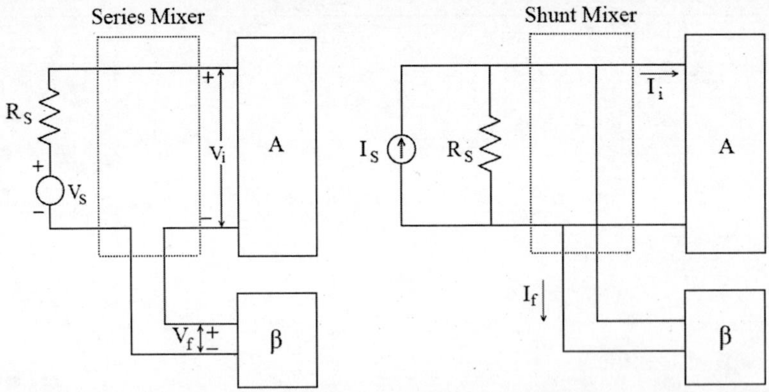

Fig. 3.2. Mixer Networks: (a) Voltage Comparator (b) Current Comparator

Sampling Network: The sampling networks are of two types:

1. Voltage sampling: The network in which the output voltage is sampled by connecting the feedback network in shunt across the output is known as voltage sampling.
2. Current sampling: The network in which the output current is sampled by connecting the feedback network in series with the output is known as current sampling.

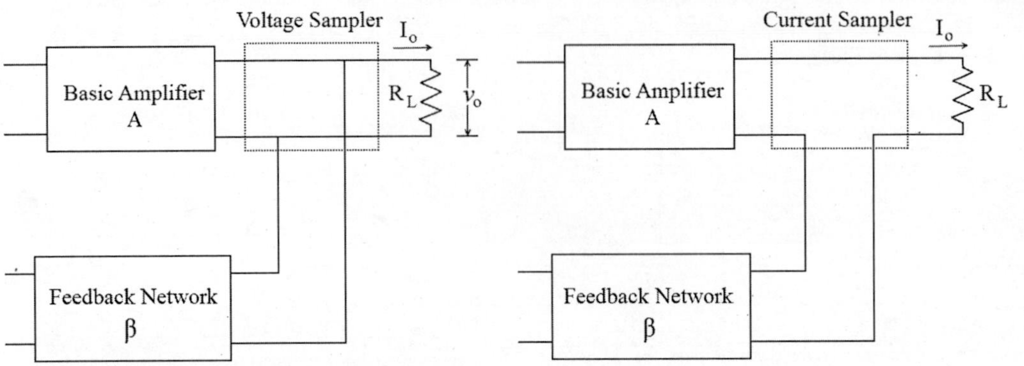

Fig 3.3. Sampling Networks: (a) Voltage Sampling (b) Current Sampling

Feedback network: It is generally a two-port network composed of passive elements such as resistors, capacitors, and inductors. Most often it is simply a resistive configuration.

Advantages of Negative Feedback:

1. Reduction in noise
2. Higher input impedance
3. Improved frequency response
4. Lower output impedance
5. Stabilized voltage gain
6. More linear operation

Disadvantages:

Negative feedback results in reduced overall voltage gain as compared to the appliers without feedback arrangements.

3.2. Feedback connection types

There are four different ways by which a feedback signal can be connected such as-

a) Voltage-series feedback
b) Current- series feedback
c) Current-shunt feedback
d) Voltage-shunt feedback

In the list above, voltage refers to connecting the output voltage as input to the feedback network; current refers to tapping off some output current through the feedback network. Series refers to connecting the feedback signal in series with the input signal voltage; shunt refers to connecting the feedback signal in shunt (parallel) with an input current source.

Series feedback connections tend to increase the input resistance, while shunt feedback connections tend to decrease the input resistance. Voltage feedback tends to decrease the output impedance, while current feedback tends to increase the output impedance. Typically, higher input and lower output impedances are desired for most cascade amplifiers. Both of these are provided using the voltage-series feedback connection.

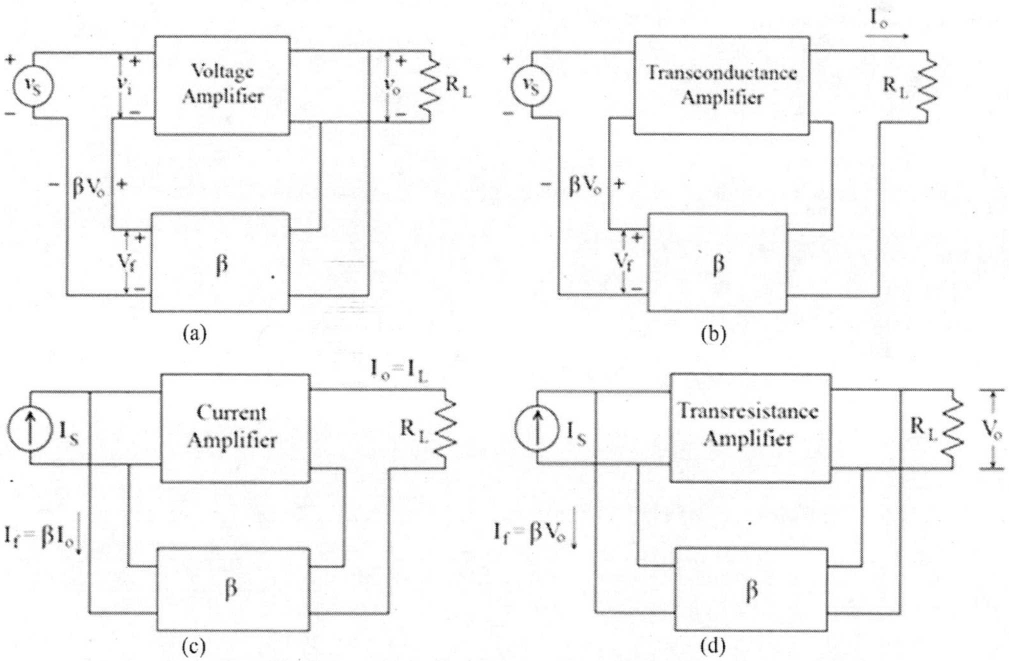

Fig 3.4. Feedback amplifiers connection type: (a) Voltage series feedback b) Current series feedback (c) Current-shunt feedback (d) Voltage-shunt feedback

3.3. The gain with feedback

The gain of each of the feedback circuit connections of Fig. 3.4. is examined in this section. The gain without feedback, *A*, is that of the amplifier stage. With feedback β, the overall gain of the circuit is reduced by a factor (1+βA), as given Table 3.1 provides the summary of the gain, feedback factor, and gain with feedback of Fig. 3.4.

Voltage-Series feedback:

Figure 3.4a shows the voltage-series feedback connection with a part of the output voltage fed back in series with the input signal, resulting in an overall gain reduction. If there is no feedback ($Vf = 0$), the voltage gain of the amplifier stage is-

$$A = \frac{V_o}{V_s} = \frac{V_o}{V_s} \quad (1)$$

If a feedback signal, V_f, is connected in series with the input, then

$$V_i = V_s - V_f$$

Since $$V_o = AV_i = A(V_s - V_f) = AV_s - AV_f = AV_s - A(\beta V_o)$$

then $$(1 + \beta A)V_o = AV_s$$

Hence, the overall voltage gain with feedback is-

$$A_f = \frac{V_o}{V_s} = \frac{A}{1+\beta A} \quad (2)$$

Equation (2) shows that the gain *with* feedback is the amplifier gain reduced by the factor (1+β*A*). This factor will be seen also to affect input and output impedance among other circuit features.

Voltage-Shunt feedback:

The gain with feedback for the network of Fig. 3.4b is-

$$A_f = \frac{V_o}{I_s} = \frac{A\,I_i}{I_i + I_f} = \frac{A\,I_i}{I_i + \beta V_o} = \frac{A\,I_i}{I_i + \beta A\,I_i}$$

$$A_f = \frac{A}{1+\beta A} \quad (3)$$

Table 3.1.: Summary of Gain, Feedback, and Gain with feedback from Fig. 3.4.

Parameter	Voltage-Series	Voltage-Shunt	Current-Series	Current-Shunt
Gain without feedback (A)	$\frac{V_O}{V_i}$	$\frac{V_O}{I_i}$	$\frac{I_O}{V_i}$	$\frac{I_O}{I_i}$
Feedback(β)	$\frac{V_f}{V_o}$	$\frac{I_f}{V_o}$	$\frac{V_f}{I_o}$	$\frac{I_f}{I_o}$
Gain with feedback(A_f)	$\frac{V_O}{V_S}$	$\frac{V_O}{I_S}$	$\frac{I_O}{V_S}$	$\frac{I_O}{I_S}$

3.4. Input impedance with feedback

Voltage series feedback: A more detailed voltage series feedback connection is shown in Fig. 3.5. The input impedance can be calculated as-

$$I_i = \frac{V_i}{Z_i} = \frac{V_s - V_f}{Z_i} = \frac{V_s - \beta V_o}{Z_i} = \frac{V_s - \beta AV_i}{Z_i}$$

$$I_i Z_i = V_s - \beta AV_i$$

$$V_s = I_i Z_i + \beta AV_i = I_i Z_i + \beta AI_i Z_i$$

$$\boxed{Z_{if} = \frac{V_s}{I_i} = Z_i + (\beta A)Z_i = Z_i(1 + \beta A)} \quad \textbf{(4)}$$

The input impedance with series feedback is seen to be the value of the input impedance without feedback multiplied by the factor $(1 + \beta A)$ and applies to both voltage-series and current-series configurations.

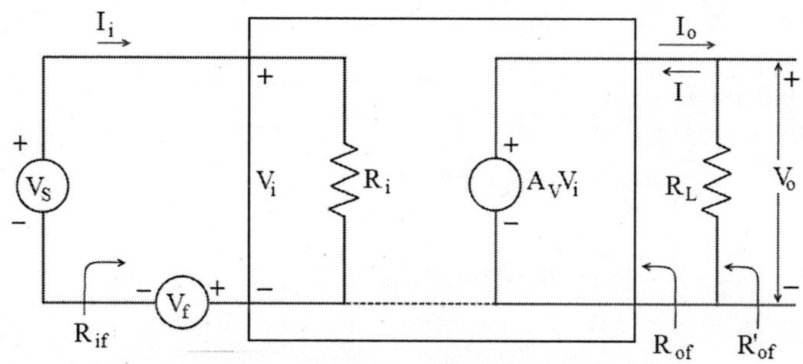

Fig. 3.5: Voltage series feedback connection

Voltage-Shunt feedback: A more detailed voltage-shunt feedback connection is shown in Fig. 3.6. The input impedance can be determined as-

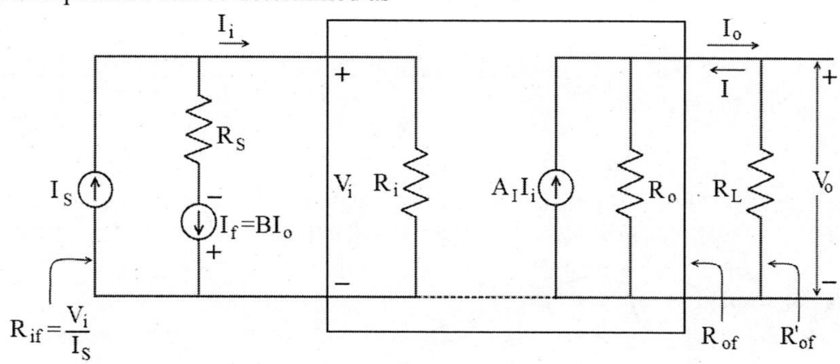

Fig. 3.6. Voltage-shunt feedback connection

$$Z_{if} = \frac{V_i}{I_s} = \frac{V_i}{I_i + I_f} = \frac{V_i}{I_i + \beta V_o}$$

$$= \frac{V_i/I_i}{I_i/I_i + \beta V_o/I_i}$$

$$\boxed{Z_{if} = \frac{Z_i}{1 + \beta A}} \qquad \textbf{(5)}$$

This reduced input impedance applies to the voltage-series connection of Fig. 3.4a and the voltage-shunt connection of Fig. 3.4b.

3.5. Output Impedance with feedback

The output impedance for the connections of Fig. 3.4 is dependent on whether voltage or current feedback is used. For voltage feedback, the output impedance is decreased, while current feedback increases the output impedance.

Voltage-Series Feedback: The voltage-series feedback circuit of Fig. 3.5 provides sufficient circuit detail
to determine the output impedance with feedback. The output impedance is determined by applying a voltage, *V*, resulting in a current, *I*, with V_s shorted out (V_s =0). The voltage *V* is then-

$$V = IZ_o + AV_i$$

For $V_s = 0$, $\quad V_i = -V_f$

so that $\quad V = IZ_o - AV_f = IZ_o - A(\beta V)$

Rewriting the equation as

$$V + \beta AV = IZ_o$$

allows solving for the output resistance with feedback:

$$\boxed{Z_{of} = \frac{V}{I} = \frac{Z_o}{1 + \beta A}} \qquad \textbf{(6)}$$

Equation (6) shows that with voltage-series feedback the output impedance is reduced from that without feedback by the factor (1+β*A*).

Current-Series Feedback: The output impedance with current-series feedback can be determined by applying a signal *V* to the output with V_s shorted out, resulting in a current *I*, the ratio of *V* to *I* being the output impedance. Fig. 3.7 shows a more detailed connection with current-series feedback. For the output part of a current-series feedback connection shown in Fig. 3.7, the resulting output impedance is determined as follows. With $V_s = 0$,

$$V_i = V_f$$

$$I = \frac{V}{Z_o} - AV_i = \frac{V}{Z_o} - AV_f = \frac{V}{Z_o} - A\beta I$$

$$Z_o(1 + \beta A)I = V$$

$$\boxed{Z_{of} = \frac{V}{I} = Z_o(1 + \beta A)} \qquad (7)$$

Table 3.2 summarizes the effect of feedback on input and output impedance.

Parameter	Voltage-Series	Current-series	Voltage-Shunt	Current-shunt
Z_{if}	$Z_i(1+\beta A)$ (increased)	$Z_i(1+\beta A)$ (increased)	$\frac{Z_i}{(1+\beta A)}$ (decreased)	$\frac{Z_i}{(1+\beta A)}$ (decreased)
Z_{of}	$\frac{Z_o}{(1+\beta A)}$ (decreased)	$Z_O(1+\beta A)$ (increased)	$\frac{Z_o}{(1+\beta A)}$ (decreased)	$Z_i(1+\beta A)$ (increased)

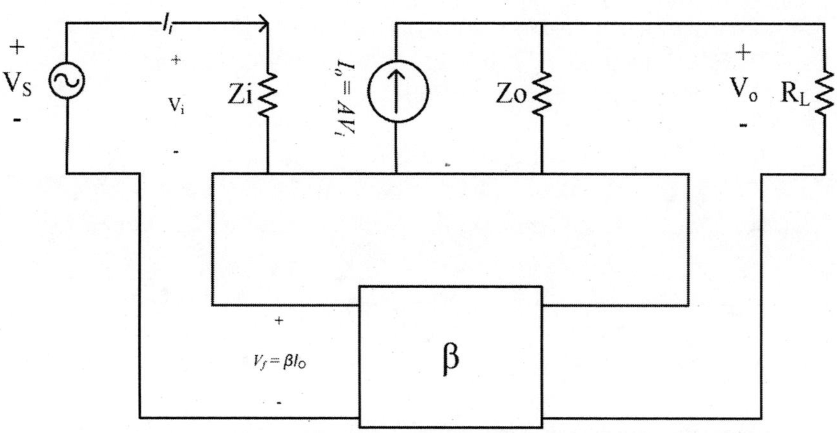

Fig. 3.7. Current-Series feedback connection

3.6. General characteristics of a negative feedback amplifier

1) Reduction in frequency distortion:

For a negative feedback amplifier, $A_f \approx \frac{1}{\beta}$ $for\ \beta A \gg 1$

It is clear from the above relation that if the feedback network is purely resistive, the gain with feedback is not dependent on its frequency even though the basic amplifier gain is frequency dependent. The frequency distortion due to varying amplifier gain is considerably reduced in negative voltage feedback amplifiers.

2) Reduction in nonlinear distortion and noise:

Signal feedback tends to reduce the noise signal and nonlinear distortion. The factor $(1 + \beta A)$ reduces both input noise and resulting nonlinear distortion. However, there is also a reduction in the overall gain of the amplifier. If additional stages are used to bring the overall gain up to the level without feedback, it should be noted that the extra stage(s) might introduce as much noise back into the system as that reduced by the feedback amplifier. This problem can be somewhat alleviated by readjusting the gain of the feedback-amplifier circuit to obtain a higher gain while also providing a reduced noise signal.

3) Effect of negative feedback on the gain and bandwidth :

The overall gain with negative feedback is-

$$A_f = \frac{A}{1 + A\beta} \approx \frac{A}{A\beta} = \frac{1}{\beta} \qquad for\ \beta A \gg 1$$

As long as $\beta A \gg 1$, the overall gain is approximately $\frac{1}{\beta}$. We should realize that for a practical amplifier, the open-loop gain drops off at high frequencies due to the active device and circuit capacitances. Gain may also drop off at low frequencies for capacitively coupled amplifier stages. Once the open-loop gain ***A*** drops low enough and the factor βA is no longer much larger than 1, the relation $A_f \cong \frac{1}{\beta}$ no longer holds correct. Fig.3.8, shows that the amplifier with negative feedback has more bandwidth (B_f) than the amplifier without feedback (B). The feedback amplifier has a higher upper 3-dB frequency and a smaller lower 3-dB frequency.

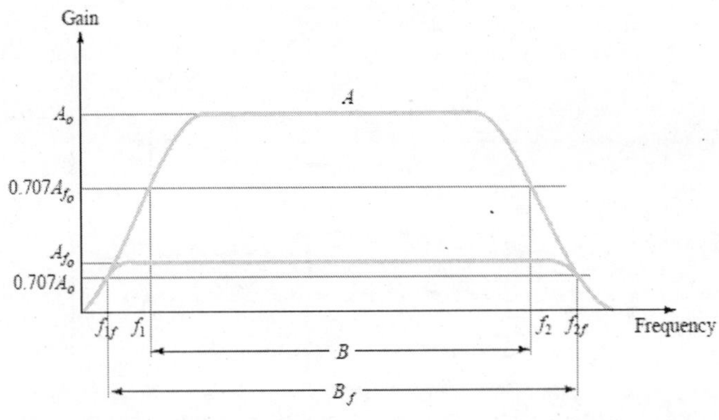

Fig. 3.8. Effect of negative feedback on gain and bandwidth

The application of a feedback signal lowers the voltage gain and increases bandwidth B (particularly in the upper 3-dB frequency). The product of gain and frequency remains the same so that the gain–bandwidth product of the basic amplifier is the same value as the feedback amplifier. However, since the feedback amplifier has a lower gain, the net operation was to *trade* gain for bandwidth.

4) Gain stability with feedback:

In addition to the β factor setting a precise gain value, the stability of a feedback amplifier is compared to an amplifier without feedback. The overall voltage gain with feedback is given by-

$$A_f = \frac{A}{1 + \beta A}$$

Differentiating the above equation with respect to A, gives-

$$\left|\frac{dA_f}{A_f}\right| = \frac{1}{|1+\beta A|}\left|\frac{dA}{A}\right|$$

$$\left|\frac{dA_f}{A_f}\right| = \frac{1}{|\beta A|}\left|\frac{dA}{A}\right| \quad for\ \beta A \gg 1$$

This shows that magnitude of the relative change in gain $\left|\frac{dA_f}{A_f}\right|$ is reduced by the factor $|\beta A|$ *as* compared to that without feedback $(\left|\frac{dA}{A}\right|)$.

3.7. Oscillators

A circuit that generates AC voltage is known as an oscillator. To generate AC voltage, the circuit is supplied from a DC source. Fig. 3.9 shows the basic difference between an amplifier and an oscillator.

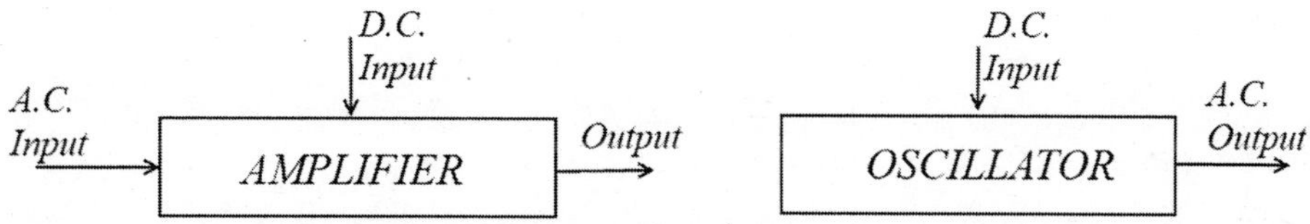

Fig. 3.9. Difference between amplifier and oscillator

An amplifier is an electric circuit, which amplifies the input signal while preserving the shape of the input waveform. A typical amplifier can have two inputs i.e. DC power input and AC signal input. The amplifier acts as an energy converter, it takes DC power input from the supply V_{CC} and converts it into AC power at the signal frequency and gives it to the load, the energy conversion is controlled by the input signal amplitude.

An oscillator is an electronic circuit that generates its output waveform. It has only one input i.e. DC power input. It converts the DC power into AC power at one signal frequency.

Principle of Oscillation:

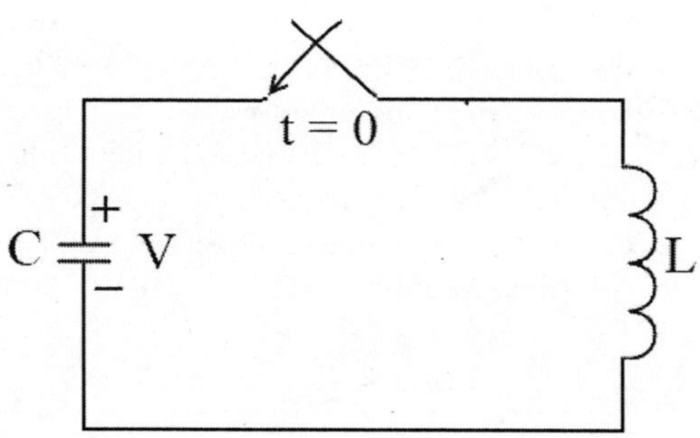

Fig.3.10: Tank circuit (LC-tuned circuit)

Consider an LC-tuned circuit as shown in Fig. 3.10. Let us assume that initially there is a charge 'Q' in the capacitor.

The voltage across capacitor $V = \frac{Q}{C}$

Let us assume an ideal capacitor. Let the ideal coil having inductance 'L' is connected across a capacitor at $t = 0$. The capacitor has got a discharge path through 'L'. Since the inductor current cannot change instantaneously, the capacitor discharges slowly. The inductor current builds up and the voltage across the capacitor decreases. In that process, the electrostatic energy in the capacitor is converted to magnetostatic energy and this is shown in Fig. 3.11 by concentric magnetic lines of force.

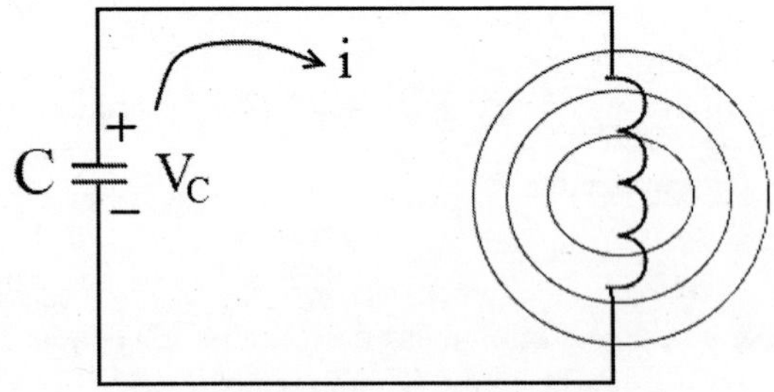

Fig. 3.11: Charging of an Inductor

When the capacitor voltage decreases to zero the current through the inductor reaches maximum. Since the capacitor cannot discharge further, the current should reduce to zero. However, the inductor opposes that change by passing the current in the same direction as shown. At that time the full electrostatic energy is converted to magneto-static energy.

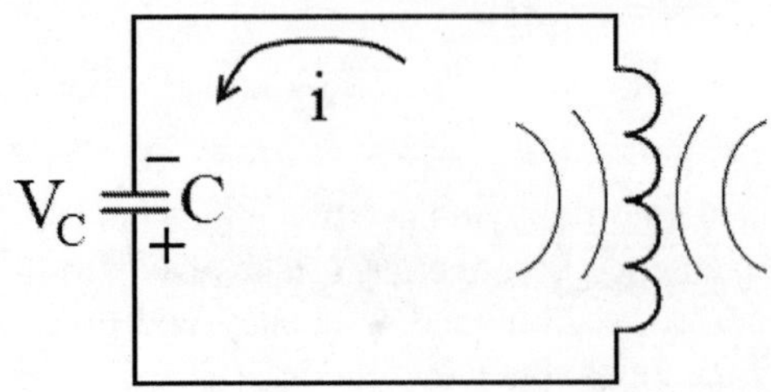

Fig.3.12: Discharging of an Inductor

When the capacitor voltage goes to zero the magnetic field around the inductor collapses as shown in Fig. 3.12 by anti-concentric lines. The capacitor with opposite polarity will charge as shown in the figure. In that process, magneto-static energy is converted into electrostatic energy. Now the capacitor takes over by discharging through an inductor in opposite direction, charging the inductor so that electrostatic energy is again converted to magneto-static energy. Later an inductor allows the flow of current in the same direction charging the capacitor. Since the inductors and capacitors are not ideal some energy will be dissipated in the series resistance of the capacitance. Hence after each cycle, the voltage across the capacitor goes on decreasing and if we observe the voltage waveform across the capacitor we get a damped sinusoidal waveform having a frequency,

$$f_0 = \frac{1}{2\pi\sqrt{LC}}$$

For un-damped oscillation following conditions should be satisfied;

a) The amount of energy supplied should be sufficient to meet the losses in the tank circuit.

b) The applied energy should have some frequency.

c) The applied energy should be in phase with oscillations set up in the tank circuit.

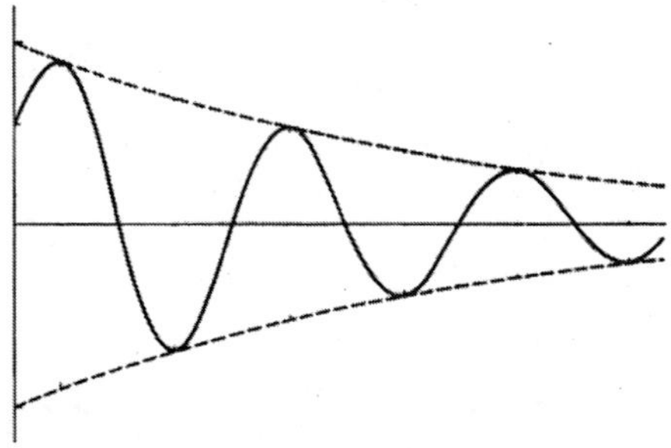

Fig.3.13: Damped Sinusoidal waveform

Due to the leakage and winding resistance, the oscillation decay after some time. By external means, the charge lost by the capacitor is supplied back. The purpose of the amplifier in the oscillator is to give back the energy lost by the storage elements.

3.8. Block Diagram of Positive Feedback Amplifier:

Fig. 3.14 shows the block diagram of the positive feedback amplifier. The feedback of the network is made up of reactive components (R, L, C). The feedback network introduces an additional phase shift such that feedback is in phase with the source voltage.

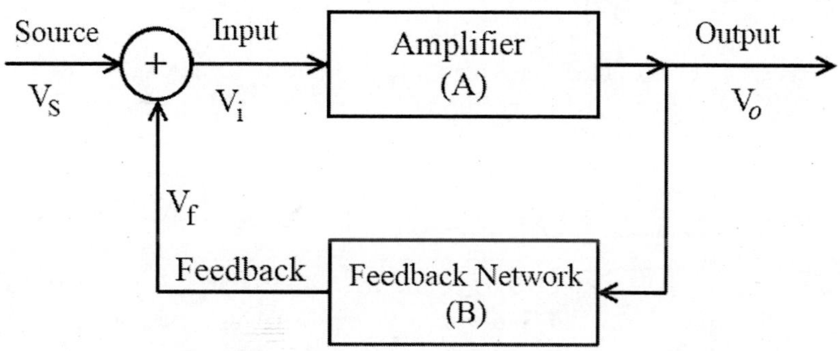

Fig. 3.14: Block diagram of positive feedback amplifier

$$\therefore V_i = V_S + \beta V_0$$

$$V_0 = A(V_S + \beta V_0) = AV_i$$

$$V_0(1 - A\beta) = AV_S$$

$$\therefore \frac{V_0}{V_S} = \frac{A}{1 - A\beta} = A_f$$

When $A\beta = 1; A_f = \infty$

i.e. gain of the amplifier with feedback is infinity.

$$\text{i.e.} \quad \frac{V_0}{V_S} = \infty$$

$$or \quad V_S = \frac{V_0}{\infty} = 0$$

i.e. In the absence of an input signal there is an output signal. So an oscillator produces its output waveform. The conditions that should be satisfied for a sustained oscillator are:

1. The loop gain should be equal to unity. i.e.

$$A\beta = 1$$

2. The loop phase shift should be zero. i.e.

$$\angle A\beta = 0, \qquad 2n\pi$$

where $n = 0,1,2,3, \ldots..$

The above two conditions are known as ***Barkhuasen Criterion***.

3.9. Classification of Oscillator

Fig. 3.15 depicts the classification chart of oscillators. An oscillator circuit producing harmonic frequencies is called a harmonic oscillator. The output of a harmonic oscillator is a pure sinusoidal waveform. Hence these oscillators are also known as sinusoidal oscillators. They are mainly classified into two types as-

1. RC oscillator
2. LC oscillator

Due to practical limitations of component ratings, RC oscillators are limited to low-frequency applications; whereas LC oscillators are used for high-frequency applications. In RC oscillators, the frequency-selective networks are made up of RC components. The RC oscillators are of two types as-

1. RC Phase Shift Oscillator
2. Wien Bridge Oscillator

An oscillator generating a non-sinusoidal waveform such as square, triangular, etc is called a relaxation oscillator. Here the circuit employs an active device operated in the non-linear region of its characteristics. It works on the principle of charging and discharging the capacitor and inductor. It is mainly divided into two types as-

1. Multi-vibrator
2. Sweep Generator

A circuit that employs devices exhibiting negative resistance properties is called a negative resistance oscillator. These oscillators use devices like UJT or tunnel diode.

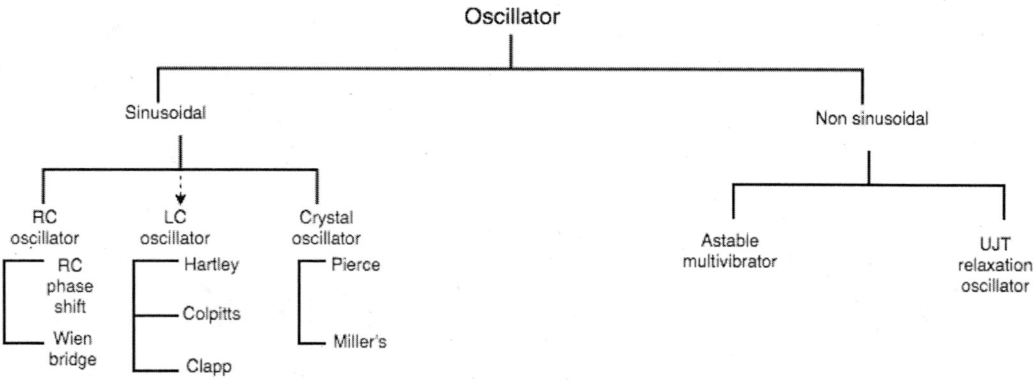

Fig. 3.15: Oscillator classification chart

3.10. RC phase shift oscillator

Fig. 3.16 shows the circuit diagram of an RC phase shift oscillator. It consists of a common emitter amplifier biased in the active region by using R_1, R_2 and R_E. Usually, the Q point is selected at the center of the DC load line. The output of the amplifier is given to a frequency selective feedback network, made up of RC components that attenuate output signal and phase shifted by 180^0.

The RC network used is a phase lead network using a single RC section, a maximum of 90° phase shifts can be realized if RC elements are ideal. Therefore two RC sections can produce 180^0 phase shifts.

Since R and C are not ideal, 90^0 phase shifts cannot be obtained by using a single RC network. Hence, a minimum of three RC networks are used. The phase shift can be adjusted by adjusting the value of R and C. We can assume that each section introduces a phase shift of 60^0. Hence, feedback produces 60×3 i.e. 180° phase shift. The amplifier introduces a phase shift of 180^0. Since it is a common emitter type, therefore total loop phase shift = 360^0 or $\angle A\beta = 0$.

By classical circuit analysis,

$$f_0 = \frac{1}{2\pi RC\sqrt{6}}$$

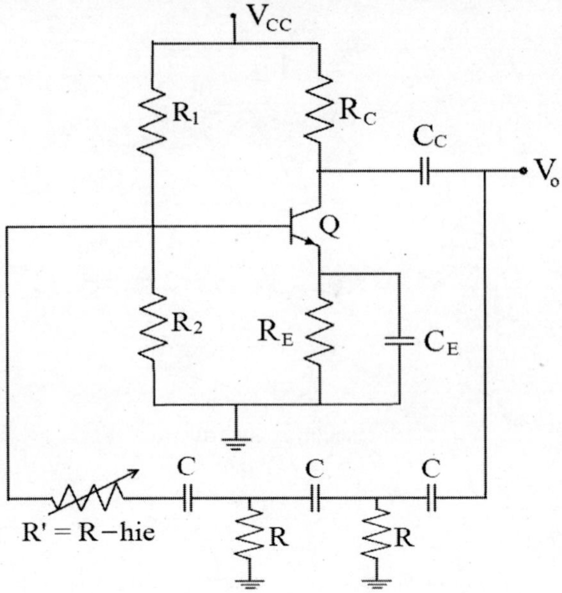

Fig. 3.16: RC phase shift oscillator

By adjusting the gain of the amplifier, the condition $A\beta \geq 1$ can be observed. Since the Barkhausen criterion is satisfied, a sustained oscillation at the output can be observed. The frequency of oscillation is given by

$$f_0 = \frac{1}{2\pi RC\sqrt{6 + 4\frac{R_C}{R}}}$$

and the condition of oscillation is given sustained oscillation is $hfe_{min} \geq 44.5$ i.e. the transistor selected should have a current gain greater than 44.5, then only oscillations build up.

The RC phase shift oscillator is used for low-frequency applications only (1Hz to 100 KHz) since $f\alpha\left(\frac{1}{RC}\right)$ at high frequencies, the value of C is very small so that it is comparable with the internal capacitance of the transistor. Hence, effective capacitance changes, changing the total phase shift around the close loop. Therefore, $\angle AB \neq 0$. Hence the oscillations cannot be sustained. If R is decreased to zero, leakage is possible.

Even when oscillators are sustained, the output frequency will not be the same as that of the desired value. The transit time effects inside the transistor are also higher. Hence CE transistor cannot produce 180^0 phase shift, therefore oscillations cannot sustain. Hence RC phase shift oscillator cannot be used for low-frequency applications.

Further, RC phase shift oscillators cannot be used for variable frequency operation as, $f\alpha\left(\frac{1}{RC}\right)$ to vary the frequency and the values of R and C need to be varied. Since three resistors and three capacitances are involved, gang tuning is not possible because three resistances and three capacitances cannot be varied simultaneously. Therefore RC Phase Shift oscillator is used for fixed-frequency applications.

Advantages:

a) It does not require a transformer or inductor

b) It can be used to provide very low frequencies

c) The circuit provides good frequency stability

Disadvantages:

a) It is difficult for the circuit to start oscillation as feedback is generally small.

b) The circuit gives small output.

3.11. Wien Bridge Oscillator:

The Wien bridge oscillator is a standard circuit for generating low frequencies in the range of 10Hz to 1MHz. It is used in all commercial audio generators. This oscillator consists of two stages of RC coupled amplifier and a feedback network. The practical circuit of the Wien bridge oscillator is shown in Fig. 3.17.

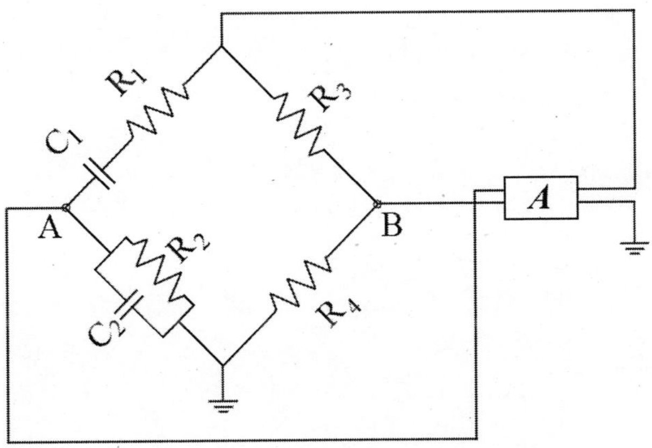

Fig. 3.17: A Wien Bridge Oscillator

Block A represents an amplifier stage. The amplifiers may be BJT or FET. The output of the amplifier goes to the feedback network. The voltage across the parallel combination $R_2\ C_2$ is fed into the input of the amplifier stage. Hence, the net phase shift through the amplifier is zero. Therefore, it is evident that for oscillation to be maintained, phase shift through the coupling network must be zero. This condition occurs at the frequency given by-

$$f_0 = \frac{1}{2\pi\sqrt{R_1 C_1 R_2 C_2}}$$

If $R=R_1=R_2$ and $C=C_1=C_2$, then

$$f_0 = \frac{1}{2\pi RC} \quad and \quad \frac{R_3}{R_4} = 2$$

The ratio of R_3 to R_4 greater than 2 will provide sufficient loop gain for the circuit toscillate at the frequency f_0. The bridge can be balanced by pulling the potentials of terminals A and B to some value. An ideal device for equating this potential of terminals A and B to the same value is an amplifier with a very high input impedance. Hence, an operational amplifier can be used.

Amplitude stabilization: Amplitude stability can be achieved by replacing the resistor R_3 with a thermistor. If the amplitude of oscillation increases, the current flowing through the thermistor increases, thereby raising its temperature and varying its resistance value. This change in resistance is used to bring down the voltage of the amplifier thereby reducing the amplitude of oscillation to a specified value. If amplitude decreases reverse process takes place.

Advantages:

a) It provides constant output;
b) The overall gain is high because of transistors;
c) The frequency of oscillations can be easily varied by using a potentiometer.
d) Amplitude stability can be achieved by using a thermistor.

Disadvantages:

- It cannot generate very high frequencies.

3.12 LC Oscillators

LC oscillators or resonant oscillators are widely used for generating high frequencies. With practical values of inductors and capacitors, it is possible to produce frequencies as high as 500MHz. The common applications are in RF generators, radio and TV receivers, high-frequency heating, etc.

3.12.1 Hartley Oscillator

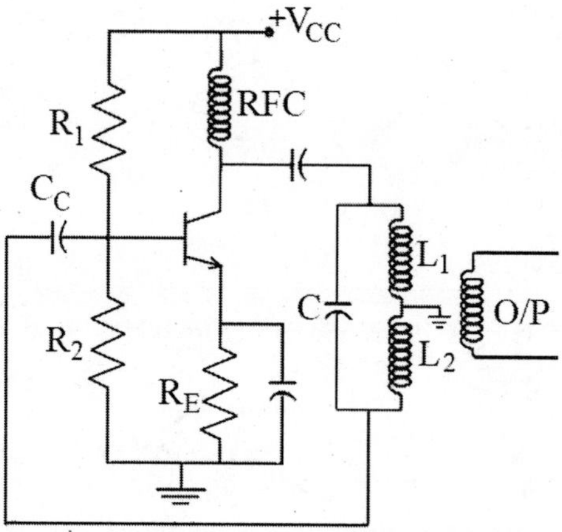

Fig. 3.18: Circuit diagram of Hartley oscillator

It consists of a positive feedback network formed by the inductance L_1L_2 and capacitance C and class A amplifier biased by R_1R_2 resistors. The function of the output voltage across L_1 is fed back to input through C_C. The tapped coil produces a correct phase shift of 180^0 for the feedback network. Resistors R_1, R_2 and R_E forms biasing circuit and RFC (Radio Frequency Choke) prevents current from passing through the power supply. From the AC point of view, RFC acts as an open circuit, and hence loading at the collector is avoided.

$$f_0 = \frac{1}{2\pi\sqrt{LC}} \quad \text{where } L = L_1 + L_2$$

$$hfe \geq \frac{L_1}{L_2} \qquad hfe \rightarrow \text{current gain of the transistor}$$

3.12.2 Colpitts Oscillator

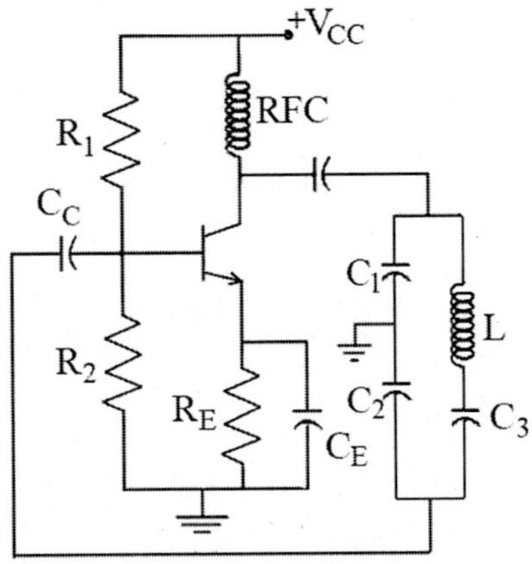

Fig. 3.19: Circuit diagram of Colpitts oscillator

When a switch is closed capacitors C_1 and C_2 are charged. These capacitors discharge through coil L, setting up oscillations of frequency given by

$$f_0 = \frac{1}{(2\pi\sqrt{LC})}$$

$$C = \frac{(C_1 C_2)}{C_1 + C_2}$$

The oscillation across C_2 are applied to the base-emitter junction and appear in amplified form in the collector circuit and supply losses to the tank circuit. The amount of feedback depends upon the relative value of C_1, C_2. 180^0 phase shift is provided by the tank circuit and another 180^0 by the amplifier and hence total phase shift is 360^0. The condition imposed on the gain of the transistor C_1, C_2 is

$$hfe > \frac{C_2}{C_1}$$

Here coupling capacitors are used to block DC produced at the collector to the tank circuit.

3.12.3 Crystal Oscillator

A Crystal oscillator is a tuned circuit oscillator using a piezoelectric crystal as a resonant tank circuit. The crystal has greater stability in holding constant at whatever frequency the crystal is originally cut to operate. Crystal oscillators are widely used in communication transmitters and receivers.

Electrical Equivalent circuit of crystal:

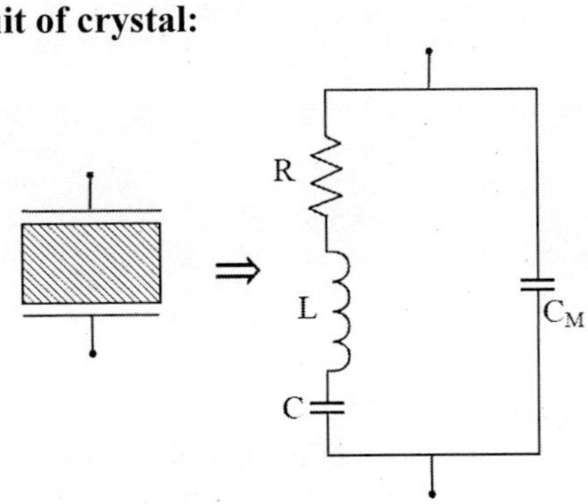

Fig. 3.20: Electrical equivalent circuit of crystal

Here inductor L and Capacitor C represent electrical equivalents of crystal mass and compliance. R is the electrical equivalent of a crystal structure's internal friction. C_M is capacitance due to the mechanical mounting of the crystal.

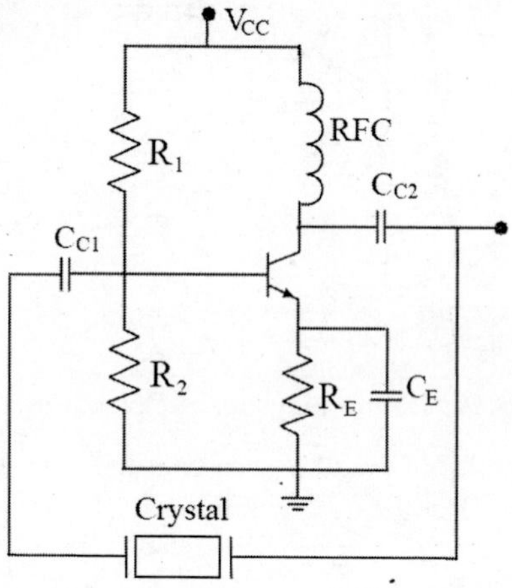

Fig. 3.21: Transistorized crystal oscillator

In this circuit, the crystal operates in series resonant mode as it offers a low impedance path for feedback signal from output to input at this frequency. Hence oscillation builds up at this frequency which is indeed f_s.

$$f_s = \frac{1}{2\pi\sqrt{LC}}$$

In Crystal Oscillators-

$$f = \frac{K}{t} \text{ where t = thickness and K = constant}$$

For high frequency should be small and extremely thin crystals may break because of the vibrations. Hence it cannot be used for high frequency. For low frequency 't' should be more which increases the size and also affects capacitance, hence cannot be used for low frequencies. f_p is always greater than f_s. The actual frequency of operation lies between f_p and f_s. Hence, a crystal oscillator cannot be used for frequencies less than 15KHz and above 10MHz. The frequency of oscillation can be controlled by any one of the following methods:

1. **Inductive Tuning:** In this case, tuning is performed by varying the inductance of the tuning inductor.
2. **Capacitive Tuning:** In this case, tuning is performed by varying the capacitance of the tuning capacitor.
3. **Mixed tuning:** In this case, tuning is performed by varying the inductance of the tuning inductor at the low-frequency end and the capacitance of the tuning capacitor at the high-frequency end.

Solved Examples

Ex. 1. Determine the voltage gain, input, and output impedance with feedback for voltage series feedback having A=-100, Ri = 10 kΩ, Ro = 20 kΩ for feedback of (a) β = -0.1 and (b) β = -0.5.

Solution:

$$\text{(a)}\ A_f = \frac{A}{1+\beta A} = \frac{-100}{1+(-0.1)(-100)} = \frac{-100}{11} = \mathbf{-9.09}$$

$$Z_{if} = Z_i\,(1+\beta A) = 10\ \text{k}\Omega\ (11) = \mathbf{110\ k\Omega}$$

$$Z_{of} = \frac{Z_o}{1+\beta A} = \frac{20\times10^3}{11} = \mathbf{1.82\ k\Omega}$$

$$\text{(b)}\ A_f = \frac{A}{1+\beta A} = \frac{-100}{1+(0.5)(100)} = \frac{-100}{51} = \mathbf{-1.96}$$

$$Z_{if} = Z_i\,(1+\beta A) = 10\ \text{k}\Omega\ (51) = \mathbf{510\ k\Omega}$$

$$Z_{of} = \frac{Z_o}{1+\beta A} = \frac{20\times10^3}{51} = \mathbf{392.16\ \Omega}$$

Ex. 2: If an amplifier with a gain of -1000 and feedback of β= -0.1 has a gain change of 20% due to temperature, calculate the change in gain of the feedback amplifier.

Solution:

$$\left|\frac{dA_f}{A_f}\right| \cong \left|\frac{1}{\beta A}\right| \left|\frac{dA}{A}\right| = \left|\frac{1}{-0.1(-1000)}(20\%)\right| = \mathbf{0.2\%}$$

The improvement is 100 times. Thus, while the amplifier gain changes from $|A| = 1000$ by 20%, the gain with feedback changes from $|A_f| = 100$ by only 0.2%.

Ex. 3: It is desired to design a phase-shift oscillator using a FET having $g_m = 5000\mu S$, $r_d = 40\ k\Omega$, and feedback circuit value of $R = 10\ k\Omega$. Calculate the value of C for oscillator operation at 1 kHz and R_D for $A > 29$ to ensure oscillator action.

$$C = \frac{1}{2\pi R f\sqrt{6}} = \frac{1}{(6.28)(10\times10^3)(1\times10^3)(2.45)} = \mathbf{6.5\ nF}$$

Ex. 4: Calculate the resonant frequency of the Wien bridge oscillator of Fig.3.22.

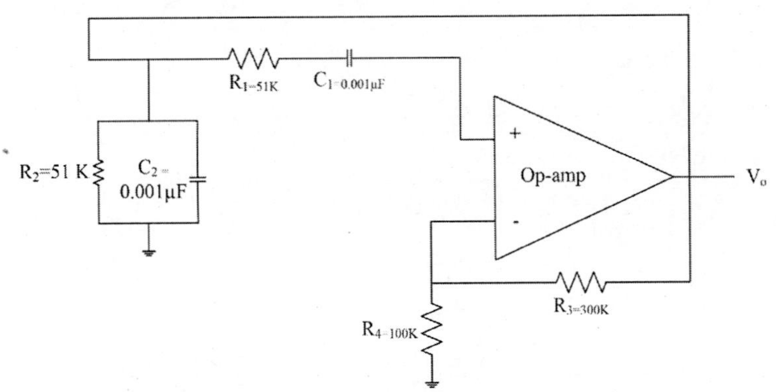

Fig. 3.22: Wien Bridge oscillator circuit

Solution:

$$f_o = \frac{1}{2\pi RC} = \frac{1}{2\pi\,(51 \times 10^{3})(0.001 \times 10^{-6})} = \mathbf{3120.7\ Hz}$$

Exercise Questions

1. With a suitable block diagram, explain different types of negative feedback circuits.
2. With neat diagrams explain different types of feedback and find their Ri, Ro, Av, and Ai
3. Explain the various feedback topologies.
4. What are the advantages of negative feedback?
5. What are the effects of negative feedback on various parameters of an amplifier?
6. Explain the principle of operation of the oscillator.
7. Discuss the Barkhausen criteria for an oscillator.
8. Draw the circuit diagram of the RC phase shift BJT oscillator and explain the working.
9. Draw the diagram of the Wien bridge oscillator. Derive the expression for resonant frequency.
10. Compare the LC and RC oscillators.
11. Draw the circuit diagram for the Hartley oscillator. Write down the expression for its frequency of oscillations.
12. Draw and explain a Colpitts oscillator.

Self-Study Questions

1. Draw a current amplifier. State its characteristics and introduce negative feedback to it and name the topology. Draw a transistorized circuit to implement the topology, what will be the important characteristics of the circuit?
2. What are the advantages of crystal oscillators over other high and low-frequency oscillators?
3. Compare the Hartley and Colpitts oscillator.
4. Differentiate between an oscillator and an amplifier.
5. Write a note on tuned oscillators.
6. Compare and contrast the positive and negative feedback systems.

Multiple Choice Questions

1. Which of the following improvements is (are) a result of the negative feedback in a circuit?
    a) Higher input impedance
    b) Better stabilized voltage gain

c) Improved frequency response
d) all of above

2. What is the ratio of the input impedance with series feedback to that without feedback?

a) $1 + \beta A$
b) βA
c) β
d) 1

3. Determine the voltage gain with feedback for voltage-series feedback having $A = -100$, $R1 = 15\ k\Omega$, $Ro = 20\ k\Omega$, and a feedback of $\beta = -0.25$.

a) 3.85
b) –3.85
c) –9.09
d) 9.09

4. At what phase shift is the magnitude of βA at its maximum in the Nyquist plot?

a) 90°
b) 180°
c) 270°
d) 0°

5. Which of the following is required for oscillation?

a) $\beta A > 1$
b) The phase shift around the feedback network must be 180°.
c) Both $\beta A > 1$ and the phase shift around the feedback network must be 180°.
d) None of the above

6. Only the condition βA = _____ must be satisfied for self-sustained oscillations to result.

a) 0
b) –1
c) 1
d) None of the above

7. In the RC phase-shift oscillator, what should the ratio of feedback resistor Rf to R1 be?

a) Zero
b) Greater than –29
c) Less than 29
d) Any value

8. For a phase-shift oscillator, the gain of the amplifier stage must be greater than ________.

a) 19
b) 29
c) 30
d) 1

9. In the Wien bridge oscillator, which of the following is (are) frequency-determining components?

a) Rl and R2
b) Cl and C2
c) Rl, R2, Cl, and C2
d) None of the above

10. Which of the following oscillators is (are) tuned oscillators?

a) Colpitts
b) Hartley
c) Crystal
d) All of the above

11. Voltage shunt feedback amplifier forms…..
a) A negative feedback
b) A positive feedback
c) Both positive and negative
d) None of the mentioned

12. Voltage shunt feedback amplifiers are also called as…..
a) Non-inverting amplifier with feedback
b) Non-inverting amplifier without feedback
c) Inverting amplifier with feedback
d) Inverting amplifier without feedback

13. In a negative feedback amplifier, voltage sampling…..
a) Tends to decrease the output resistance
b) Tends to increase the output resistance
c) Does not alter the output resistance
d) Produces the same effect on output resistance as current sampling

14. In a negative feedback amplifier, shunt mixing:
a) Tends to increase the input resistance
b) Tends to decrease the input resistance
c) Does not alter the input resistance
d) Produces the same effect on input resistance as the series mixing

15. An amplifier with resistive negative feedback has two left-half plane poles in its open-loop transfer junction. The amplifier will be:

a) Unstable for all frequency
b) Stable for all frequencies
c) stable for selective frequencies
d) None of above

16. Barkhausen criteria is:
a) Positive feedback, $A\beta = 1$, $\theta = 0$ or multiple 360
b) Negative feedback, $A\beta = 1$, $\theta = 0$ or multiple 360
c) Positive feedback, $A\beta = 0$, $\theta = 0$ or multiple 360
d) Negative feedback, $A\beta = 1$, $\theta = 180$

MCQ Answer key:

1	**d**	5	**c**	9	**c**	13	**b**
2	**a**	6	**c**	10	**d**	14	**b**
3	**b**	7	**b**	11	**a**	15	**b**
4	**b**	8	**b**	12	**c**	16	**a**

@@@@@@@@@

4. Operational Amplifier and Applications

RATIONALE

The operational amplifier (Op-Amp) traces its origin and name to analog computers, where they were used to perform mathematical operations in linear, non-linear, and frequency-dependent circuits. Op-Amps are used widely in electronic devices today, including a vast array of consumer, industrial, and scientific devices. Considering these facts, in this chapter, we will study the fundamentals of Operational Amplifiers and their applications.

UNIT OUTCOMES

U1-O1: Unit-1 Learning Outcome-1

To know about the basic block diagram and characteristics of an operational amplifier.

U1-O2: Unit-1 Learning Outcome-2

To know about the symbol of Op-Amp, the pin configuration of IC741, and the concept of virtual ground.

U1-O3: Unit-1 Learning Outcome-3

To study the different applications and their circuitry using Op-Amp.

LEARNING OBJECTIVES

LO1: To understand the concept of an Op-Amp.

LO2: To study the basic characteristics of an Op-Amp.

LO3: To study the concept of virtual ground.

LO4: To study the open loop configurations of an Op-Amp

LO5: To study the close loop configurations of an Op-Amp

LO6: To study the different applications of an Op-Amp.

MAPPING THE UNIT OUTCOMES WITH THE COURSE OUTCOMES

Unit Outcome	EXPECTED MAPPING WITH COURSE OUTCOMES (1- Weak Correlation; 2- Medium correlation; 3- Strong Correlation)						
	CO-1	**CO-2**	**CO-3**	**CO-4**	**CO-5**	**CO-6**	**CO-7**
U3-O1	**1**	--	--	--	**3**	--	--
U3-O2	**1**	--	--	--	**3**	--	--
U3-O3	**1**	--	--	--	**3**	--	--

Interesting Facts:

1. In 1947, the operational amplifier was first formally defined and named by John R. Ragazzini of Columbia University.
2. In 1953, vacuum tube op amps became commercially available with the release of the model K2-W from George A. Philbrick Researches, Incorporated.
3. By 1961, solid-state, discrete Op-Amp was being produced. These op amps were effectively small circuit boards with packages such as edge connectors.
4. An operational amplifier (often Op-Amp) is a DC-coupled high-gain electronic voltage amplifier with a differential input and, usually, a single-ended output. In this configuration, an Op-Amp amp produces an output potential (relative to circuit ground) that is typically 100,000 times larger than the potential difference between its input terminals.
5. Op-Amps may be packaged as components or used as elements of more complex integrated circuits.

Video Resources:

Sr	Title	URL	QR Code
1.	Introduction to Operational Amplifier: Characteristics of Ideal Op-Amp	https://www.youtube.com/watch?v=kiiA6WTCQn0&list=PLwjK_iyK4LLDBB1E9MFbxGCEnmMMOAXOH	
2.	The Concept of Virtual Ground in Op Amp	https://www.youtube.com/watch?v=AuZ00cQ0UrE	

3.	CMRR (Common Mode Rejection Ratio)	https://www.youtube.com/watch?v=hpCu3HbAiWg	
4.	Summing Amplifier (Inverting and Non-Inverting Summing Amplifiers)	https://www.youtube.com/watch?v=jsKSfaFQ4d4	
5.	Op-Amp Differentiator	https://www.youtube.com/watch?v=aU24RWIgJVs	

4.1 General characteristics of an Op-Amp

An operational amplifier, or Op-Amp, is one of the basic building blocks of analog electronic circuits. Operational amplifiers are linear devices used widely in signal conditioning, filtering, and performing different arithmetic operations. It is a direct coupled high gain device that can be used to amplify DC as well AC voltages. Due to its capacity to perform different arithmetic operations, besides basic amplification, it is named an operational amplifier. Fig. 4.1 shows the block diagram of an Op-Amp.

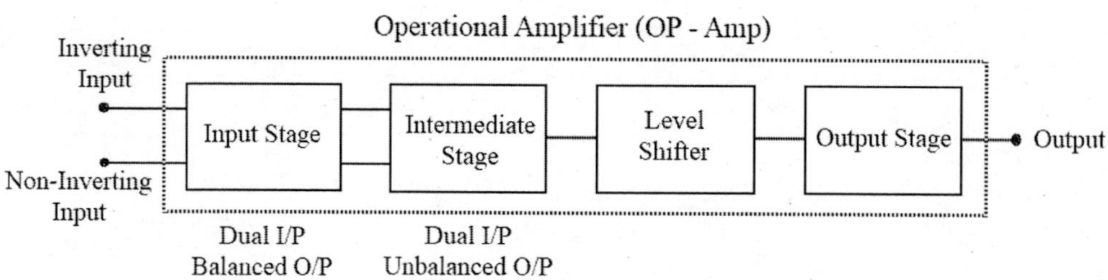

Fig. 4.1 Block diagram of Op-Amp

1. **Input stage:** It is dual input, balanced output differential amplifier. This stage provides a voltage gain of the Op-Amp and decides the input impedance. Thus this stage is used to provide high input impedance.

2. **Intermediate stage:** It is dual input unbalanced output differential amplifier. Due to high input impedance requirements, voltage gain is compromised at the first stage. Hence, the second stage must have a high voltage gain to make the overall voltage gain very high.

3. **Level Shifter:** In Op-Amp, the different stages are directly connected without using coupling capacitors. Due to it, the output DC voltage is well above ground potential; which is not desirable. The level shifter circuits help to bring the DC levels at the output to zero voltage. The simplest form of a level translator circuit is using an emitter follower with a

voltage divider. By selecting the proper values of R_1 & R_2 the voltage at the junction of R_1 & R_2 can be reduced to zero voltage.

4. **Output stage:** The final stage is usually a push-pull complementary amplifier output stage. The output stage increases the output voltage swing and raises the current-supplying capability of the Op-Amp.

Fig. 4.2 shows the symbol of Op-Amp. An ideal Operational amplifier is a three-terminal device. It consists of two high-impedance input terminals.

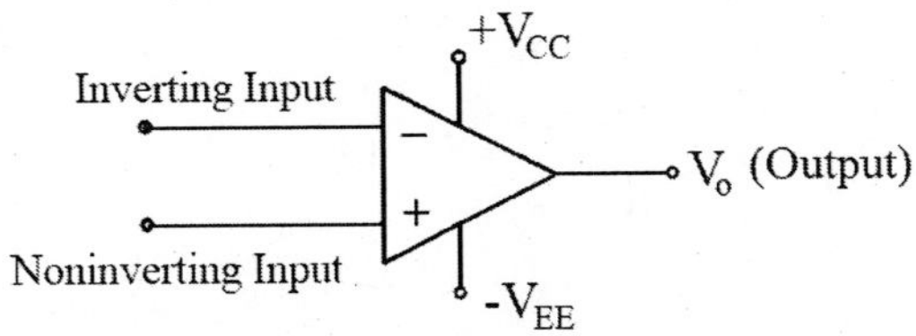

Fig. 4.2 Symbol of Op-Amp

The positive input terminal is known as the non-inverting input terminal. Any signal applied to this input terminal (with other input grounded) produces in-phase output voltage. The negative input terminal is known as the inverting input terminal as any signal applied to this input (with other input grounded) produces a 180° phase shift or a signal of opposite polarity in the output voltage.

Operational amplifiers are available in IC packages of single, dual, or quad Op-Amps within one single device. IC 741 is the most commonly available Op-Amp of the industry standard. Fig. 4.3 shows the pin configuration of a typical IC 741 Op-Amp.

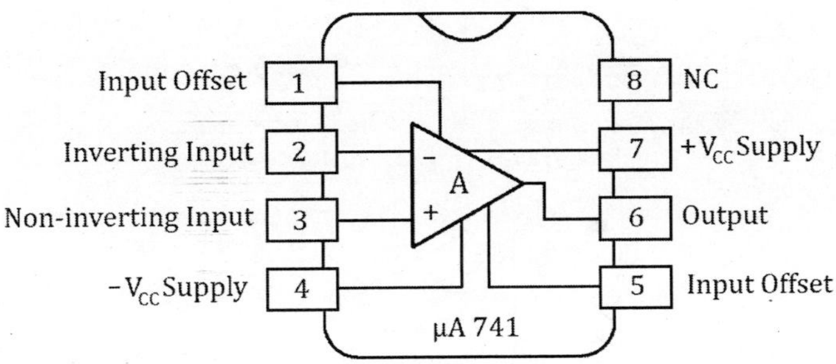

Fig 4.3: Pin configuration of IC 741 Op-Amp

Following are the few important characteristics of an ideal Op-Amp-

1. Infinite voltage gain.
2. Infinite input resistance(R_i), so that any signal source can drive it and there is no loading of the preceding stage.
3. Zero output resistance(R_o), so that output can drive an infinite number of other devices.
4. Zero output voltage when the input voltage is zero.
5. Infinite bandwidth so that any free signal from 0 to ∞ Hz can be amplified without attenuation.
6. Infinite common mode rejection ratio (CMRR), so that common mode noise voltage is zero.
7. Infinite slew rate so that output voltage changes simultaneously with input voltage change.
8. Zero Supply voltage rejection ratio (SVRR).

Fig. 4.4 shows the equivalent circuit of OP-Amp. It has an input resistance, two ends of which indicate the two terminals of Op-Amp. It acts as a voltage-dependent voltage source as indicated in the output circuit. Fig. 4.5 shows an Op-Amp in an ideal case. Under ideal conditions-

a) The input impedance is infinite - i.e. no current ever flows into either input of the Op-Amp.
b) The output impedance is zero - i.e. the Op-Amp can drive any load impedance to any voltage.
c) The open-loop gain (A) is infinite.
d) The bandwidth is infinite.
e) The output voltage is zero when the input voltage difference is zero.

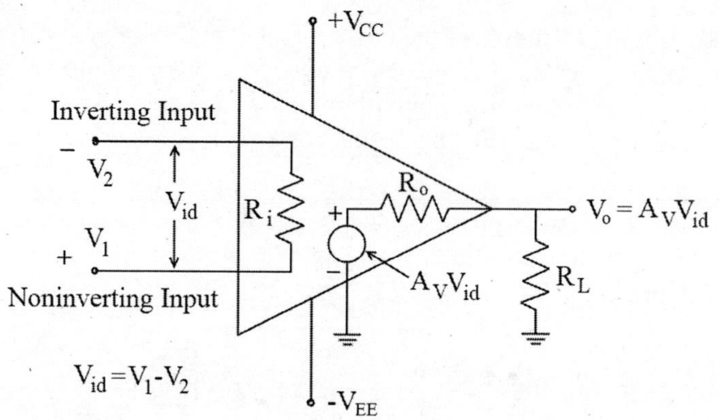

Fig.4.4: Op-Amp equivalent circuit

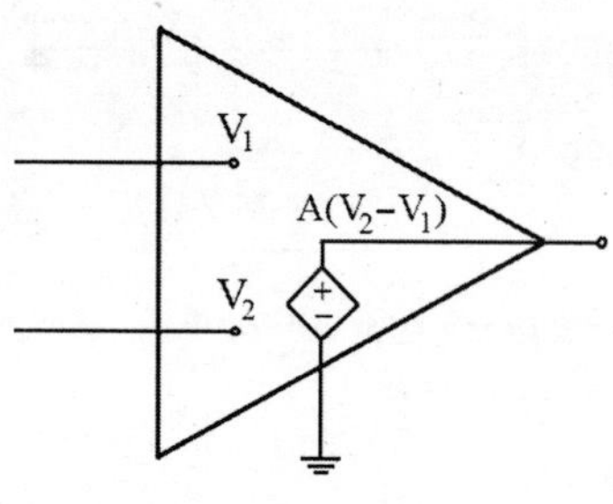

Fig.4.5: Op-Amp equivalent circuit for an ideal case

4.2. Important Specification of Op-Amp

1. Output offset voltage (V_{oo}):

The voltage obtained at the output terminals of an Op-Amp without the application of any input signal is known as output offset voltage. It is in the order of mV.

The input stage of an Op-Amp is dual input, balanced output differential amplifier. If the two transistors of the input stage are not perfectly matched, then their DC collector currents will be slightly different. Hence, between the two collectors, the output will not be zero. A DC voltage exists there without any input signal applied, which is output offset voltage.

2. **Input offset Voltage (V_{io}):**

The voltage that must be applied between the two input terminals to reduce the output offset voltage to zero, is called Input offset voltage (V_{io}). The input offset voltage may be positive or negative. The greater the matching between the inputs terminals of an Op-Amp, the lesser is input offset voltage.

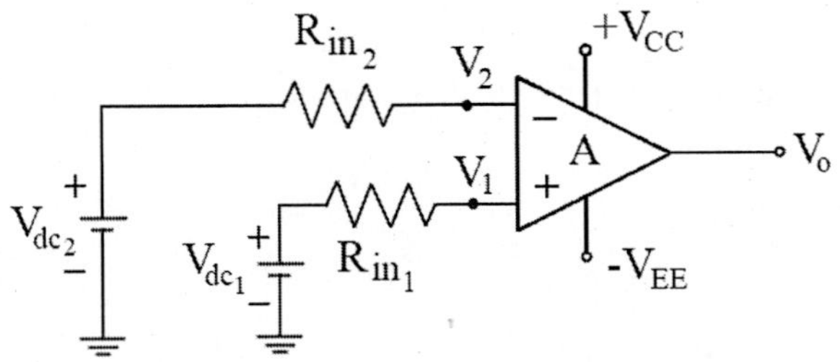

Fig. 4.6: Circuit arrangement for calculation of offset voltage

3. **Input offset current(I_{io}):** The algebraic difference between the currents into the inverting and non-inverting terminals of an Op-Amp is called input offset current I_{io}. Typically, $I_{io}\ is\ of\ order\ 200nA$.

i.e. $$I_{io} = |I_{B_1} - I_{B_2}|$$

Where I_{B_1} = current into the non-inverting input

I_{B_2} = current into the inverting input.

4. **Input Bias Current (I_B):** It is the average current that flows into the inverting and non-inverting terminals of an Op-Amp. Typically $I_B\ is\ 500\ nA$(maximum).

$$\therefore I_B = \frac{I_{B_1}+I_{B_2}}{2}$$

5. **Common Mode Rejection Ratio (CMRR):** It is the ability of an Op-Amp to reject undesired disturbances that might be amplified with the input signal. A typical value of CMMR is 90dB.

$$CMRR = \frac{|A_d|}{|A_C|}$$

A_d = Differential mode gain

A_C = Common mode gain

6. **Large signal voltage Gain (A):** It is defined as the ratio of the output voltage of an Op-Amp to the difference between the input voltages. The typical value of A is 2×10^5.

$$A = \frac{Output\ Voltage}{Differential\ Input\ Voltage}$$

$$\therefore A = \frac{V_o}{V_{id}}$$

7. **Maximum output range:** It is defined as the range of the output voltage or the value of voltages up to which the output voltage extends. For Op-Amp IC 741-

Supply voltage $V_{CC} = +15\ V,\ V_{EE} = -15V$

$\therefore$ Output voltage swing = $-13V\ to\ +13\ V.$

$$V_{opp} = 26\ V.$$

8. **Input Voltage Range:** When the same voltage is applied to both input terminals, it is called common Mode voltage(V_{CM}).

   $\therefore$ Maximum range of Input = $\pm$ 13 V

Common mode configuration is used only to determine the degree of matching between inverting and non-inverting input terminals.

9. **Supply Voltage Rejection Ratio($SVRR$):** It is the ratio of change in input offset voltage V_{io} to change in supply voltage. SVRR is of the order of 150 $\frac{\mu V}{V}$.

$$SVRR = \frac{\Delta V_{io}}{\Delta V}$$

10. **Slew Rate (SR):** It is the maximum rate of change of output voltage per unit time of an Op-Amp.

$$SR = \left.\frac{dV_o}{dt}\right|_{max}\ V/\mu\ sec \qquad \text{[Typically 0.5 } /\mu\ sec\text{ ; ideally} = \infty\text{]}.$$

11. **Gain Bandwidth Product:** It is defined as the product of the amplifier gain and bandwidth of an amplifier. It is the bandwidth of the Op-Amp when the voltage gain is unity. It is also called closed-loop Bandwidth, or unit gain Bandwidth. The typical value of GBP is 1 MHz.

4.3. Concept of Virtual ground

A virtual ground (or virtual earth) is a node of a circuit that is maintained at a steady reference potential, without being connected directly to the reference potential. An Op-Amp inverting input (-) is at zero potential (A virtual ground), even though it does not have a galvanic connection to the ground.

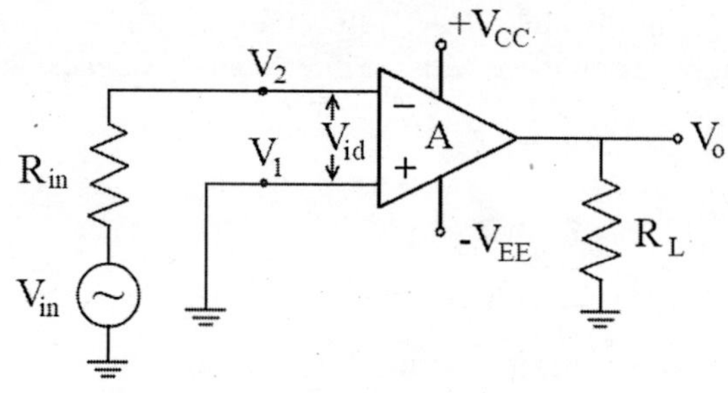

Fig.4.7. Op-Amp inverting amplifier

Here non-inverting input is grounded and voltage is applied to the inverting input. According to the virtual ground concept, the voltage at inverting node is zero.

$$\therefore V_o = -A\ V_{id}$$

$$\therefore V_o = A\ V_{id} = A(V_1 - V_2)$$

or $A = \frac{V_o}{V_{id}}$ $[V_{id} = V_1 - V_2]$

$$\Rightarrow V_{id} = \frac{V_o}{A}$$

Ideally $A = \infty$ (open loop gain)

$$\therefore V_{id} \approx 0$$

$$\therefore V_1 - V_2 \cong 0$$

But $V_1 = 0 \therefore V_2 = 0$

Hence whenever the non-inverting terminal is grounded, its inverting terminal is at the virtual ground signal.

Difference between ground and virtual ground: Ground always sinks the current and virtual ground sinks the current as well as sources the current.

4.4. Open Loop configurations of an Op-Amp

In the case of an amplifier, the term open loop indicates that there is no connection exists between output and input. When connected in an open loop configuration Op-Amp functions as a high-gain amplifier. When the output reaches saturation level, it remains constant irrespective of changes in input voltage. The output switches between positive and negative saturation in an open-loop configuration. Hence, this configuration is less popular in linear applications. Inverting amplifiers, non-inverting amplifiers, and differential amplifiers are some of the famous open-loop Op-Amp configurations.

4.4.1. Inverting Amplifier: In this configuration, only inverting input terminal is active and the non-inverting input terminal is grounded. The input is amplified by factor ***A*** *(open loop gain)* and the output is 180° out of phase with the input. The output voltage is either positive or negatively saturated. Fig. 4.8 shows an inverting amplifier using Op-Amp.

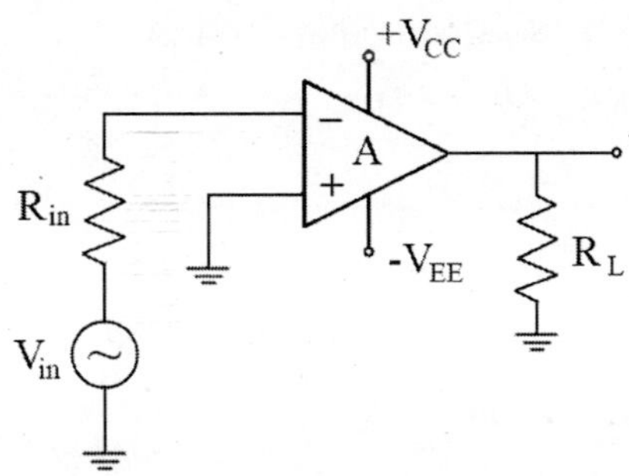

$$V_o = -A\,V_{in}$$

Fig. 4.8: Open loop inverting amplifier

4.4.2. Non-Inverting Amplifier: In this configuration, only the non-inverting input terminal is used and the inverting terminal is grounded. The input is amplified by the factor ***A*** (open loop gain) and the output is in phase with the input. The output voltage is either positive or negatively saturated. Fig. 4.9 shows the non-inverting amplifier using OP-Amp.

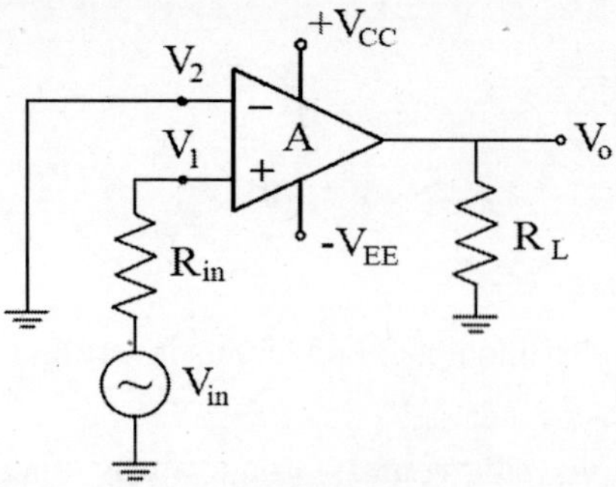

$$V_o = A\,V_{in}$$

Fig. 4.9: Open-loop non-inverting amplifier

4.4.3. Differential Amplifier: In this application, Op-Amp amplifies the difference between AC or DC input signals. Fig. 4.10 shows the typical differential amplifier using OP-Amp.

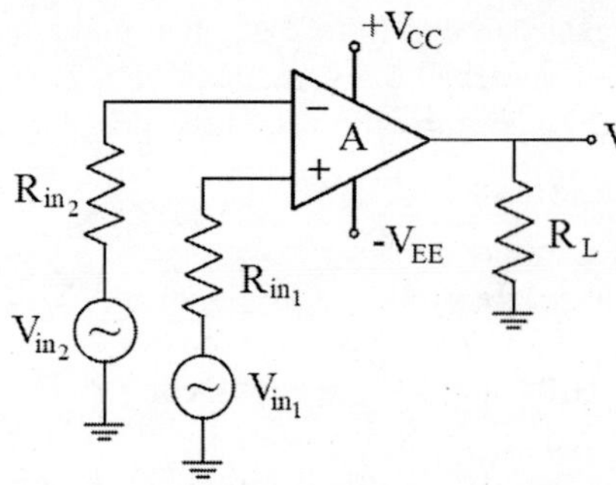

$$V_o = A(V_{in_1} - V_{in_2})$$

Fig. 4.10: Differential amplifier using OP-Amp

Polarity of V_o depends on the polarity of V_{in_1} and V_{in_2}.

$$V_o \propto V_{in_1} - V_{in_2}$$

$$\therefore V_o = AV_{id}$$

The output voltage is either positive saturated or negative saturated depending on V_{id}.

Limitations of open loop configuration in OP-Amp:

1. The output levels are fixed at $\pm V_{sat}$. When the output attempts to exceed these levels, clipping or distortion occurs. Hence very small signals having low amplitude can be amplified properly.
2. The open loop gain is not constant. It varies with changes in temperature and supply (i.e. unstable).
3. Due to a very small bandwidth, an open-loop Op-Amp configuration cannot be used for AC applications at high frequencies.

4.5. Close loop configurations of an Op-Amp

To overcome the difficulties associated with the open loop configurations, feedback is used. The resulting configuration is known as the close loop configuration of the Op-Amp. Following are some famous close-loop configurations of Op-Amp.

4.5.1. Non–Inverting Amplifier with Feedback [Voltage Series Feedback]

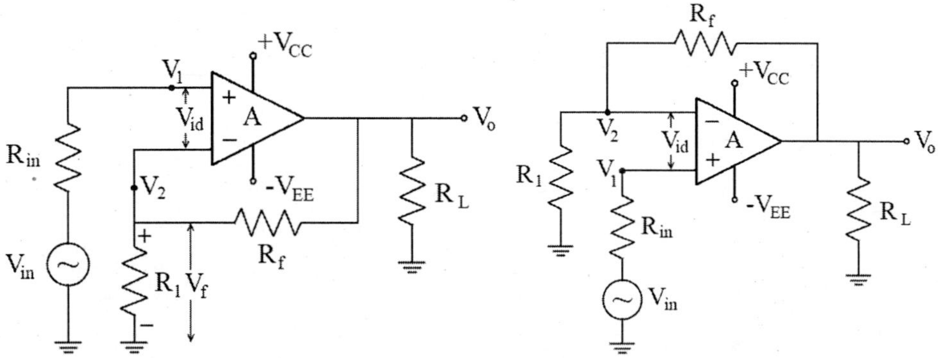

Fig. 4.11: Non–Inverting Amplifier with Feedback

Here voltage to be amplified is applied to non – inverting terminal. The output voltage is feedback in series with the input voltage. The voltage across R_1 is feedback to input.

$$V_f = \frac{V_o R_1}{R_1 + R_f}$$

$$\therefore \ V_f = \beta \, V_o$$

$$\therefore where \ \beta = \frac{R_1}{R_1 + R_f}$$

Now $V_o = A\,(V_1 - V_2)$

Where $A = open\ loop\ gain$

$$V_2 = V_o \frac{R_1}{R_1 + R_f}$$

$$\therefore V_o = AV_1 - \frac{AV_o R_1}{R_1 + R_f}$$

$$V_o \left[1 + \frac{AR_1}{R_1 + R_f}\right] = AV_1 \ \dots\dots (1)$$

$$\therefore \ \frac{V_o}{V_1} = \frac{A}{1 + A\beta}$$

$$\therefore \ A_f = \frac{A}{1 + A\beta}$$

Also from equation (1)

$$V_o \left[\frac{R_1 + R_f + AR_1}{R_1 + R_f}\right] = AV_1$$

As A is very large

$$\therefore \ AR_1 \gg R_1 + R_f$$

$$V_o \left[\frac{AR_1}{R_1 + R_f}\right] = AV_1$$

$$A_f = \frac{V_o}{V_i} = \frac{R_1 + R_f}{R_1} \cong \frac{1}{\beta}$$

Input impedance:

$$R_i = \left.\frac{V_i}{I_i}\right| V_o = 0$$

$$= \frac{V_i - V_2}{I_i}$$

Apply KVL to input

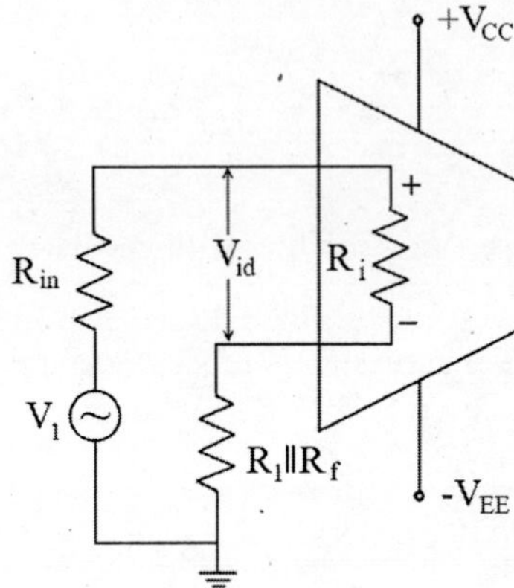

Fig. 4.12: Equivalent circuit for input impedance calculation

$$R_{inf} = \frac{V_{in}}{I_{in}}$$

$$= \frac{V_{in}}{V_{id}/R_i}$$

$$= R_i \times \frac{V_{in}}{V_{id}}$$

Now $V_{id} = V_o/A$

$$R_{if} = R_i \times \frac{V_{in}}{V_o} A$$

Now, $A_f = V_o/V_{in}$ & $A = V_o/V_{id}$

$$R_{if} = R_i \times \frac{A}{A/(1 + A\beta)} = R_i(1 + A\beta)$$

$$\therefore \; R_{if} = R_i\,(1 + A\beta)$$

Output Impedance:

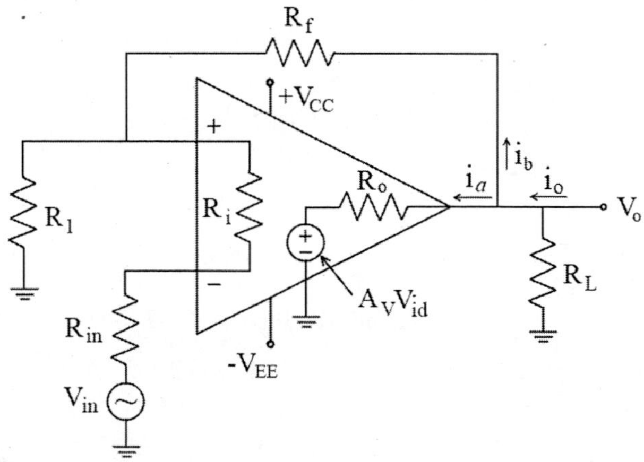

Fig. 4.13: Equivalent circuit for output impedance calculation

$$R_o = \left.\frac{V_o}{I_o}\right| V_i = 0$$

$$R_{of} = \frac{V_o}{i_o'}$$

$$i_o = i_a + i_b$$

As $R_o \cong 0$, i_b negligible

$$\therefore i_o = i_a \qquad \{ i_b \text{is negligible as compared to } i_a\}$$

$$\therefore\ i_o = \frac{V_o - A_V.V_{id}}{R_o} \ldots\ldots\ldots (1)$$

$$V_{id} = V_1 - V_2$$

and

$$V_{id} = -V_o.\frac{R_1}{R_1 + R_f}$$

$$V_{id} = -V_o.\beta$$

$$i_o = \frac{V_o(1 + A_V\beta)}{R_o}$$

Substituting in equation

$$R_{of} = \frac{V_o}{i_o} = \frac{R_o}{1 + A_V\beta}$$

4.5.2. **Inverting Amplifier with feedback** [Voltage Shunt Feedback]

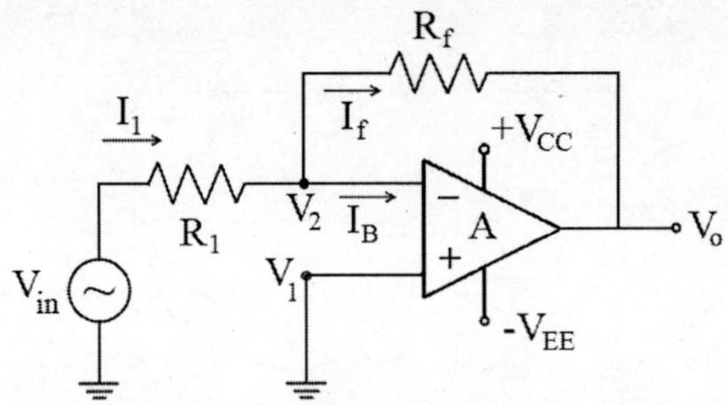

Fig. 4.14: Inverting Amplifier with feedback

As V_1 is connected to the ground and applying the virtual concept $V_2 \cong 0V$

$$I_f = -\frac{V_o}{R_f} = \beta V_o$$

$$\beta = -1/R_f$$

Apply KCL at node V_2

$$I_1 = I_f + I_B$$

$I_B \cong 0,\ \text{as } R_i \cong \infty$

$$I_1 = I_f$$

$$\frac{V_{in} - V_2}{R_1} = \frac{V_2 - V_o}{R_f}$$

As $V_2 \cong 0$ [by virtual ground concept]

$$\therefore \frac{V_{in}}{R_1} = -\frac{V_o}{R_f}$$

$$A_f = \frac{V_o}{V_{in}} = -\frac{R_f}{R_1}$$

4.5.3. Buffer Amplifier or Voltage follower

In this configuration non-inverting amplifier with unity gain is used. Therefore, the output voltage is equal to the input voltage and is in phase. Output follows the input and hence, this configuration is also called a voltage follower.

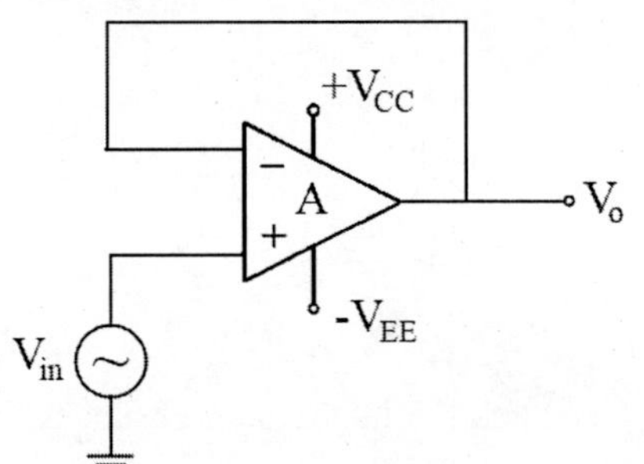

$$V_o = \left(1 + \frac{R_f}{R_1}\right) V_i$$

Fig. 4.15: Voltage follower circuit

For $V_o = V_i$, Let $R_f = 0$ and $R_1 = \infty$

Therefore, shorting the input terminal to the output and removing resistance R_1.

Now, $V_{id} = V_1 - V_2 \cong 0\ (R_i \cong \infty)$

$$\therefore V_1 - V_2 = V_{in}$$

If V_2 is equal to V_o then this circuit can be used as a voltage follower. Unlike a transistorized voltage follower, the Op-Amp voltage follower provides output exactly equal to the input. Further, almost all of the voltage from a previous source is dropped across the Op-Amp due to its characteristics of high input impedance and low output impedance. As a result, an Op-Amp can feed the rest of the circuit with the higher desired voltage.

4.5.4. Inverter (Sign Changer)

In this configuration, an Op-Amp is used as an inverting amplifier with feedback. The gain of an amplifier is given by

$$A_{vf} = -\frac{R_f}{R_1}$$

Therefore, the output voltage is

$$V_o = -\frac{R_f}{R_1} V_i$$

Let $R_f = R_1$ then $V_o = -V_i$. Hence, the sign is changed. Fig. 4.16 shows the practical circuit of an Inverter using OP-Amp.

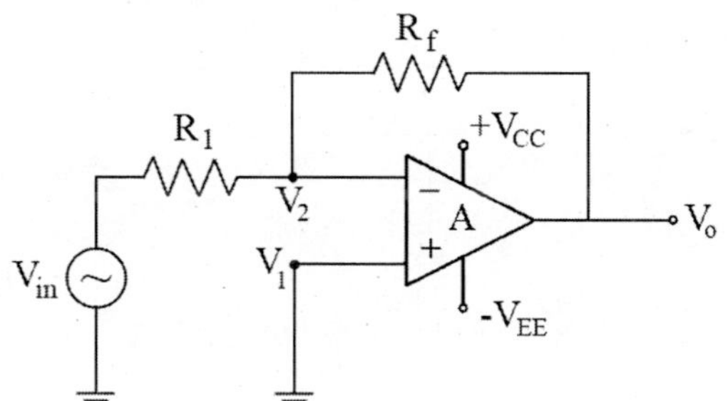

Fig. 4.16: Op-Amp as an Inverter

4.5.5. Current to voltage converter

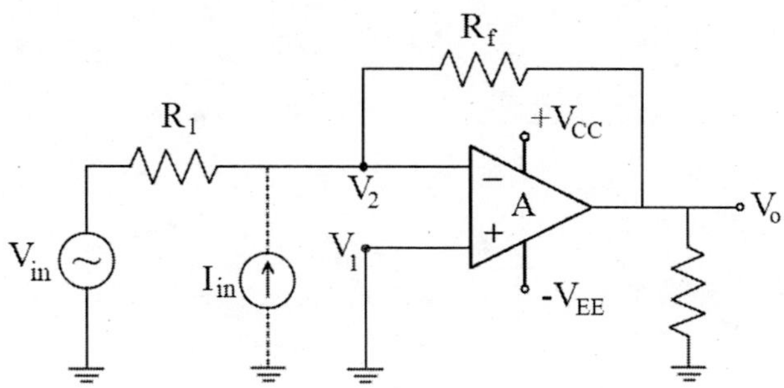

Fig. 4.17: Op-Amp as Current to voltage converter

Fig. 4.17 shows the circuit arrangement for a typical current-to-voltage converter using Op-Amp. If the combination of the voltage source V_{in} and R_1 is replaced by the current source I_{in}, the output voltage V_o becomes proportional to I_{in}.

$$I_{in} = \frac{V_{in}}{R_1}, \qquad V_o = -\frac{R_f}{R_i}.V_{in}$$

$$V_o = -I_{in}.R_f$$

4.6. APPLICATIONS OF OP-AMP

4.6.1. Summing, Scaling, and Averaging Amplifiers.

a) Inverting configurations: In this configuration, Op-Amp is used in inverting amplifier mode. There are three input V_a , V_b and V_c connected to inverting terminal. Depending on the relationship between R_f and R_a , R_b , R_c, the circuit can be used as a summing, scaling, and averaging amplifier. Fig. 4.18 shows an inverting summing amplifier using Op-Amp.

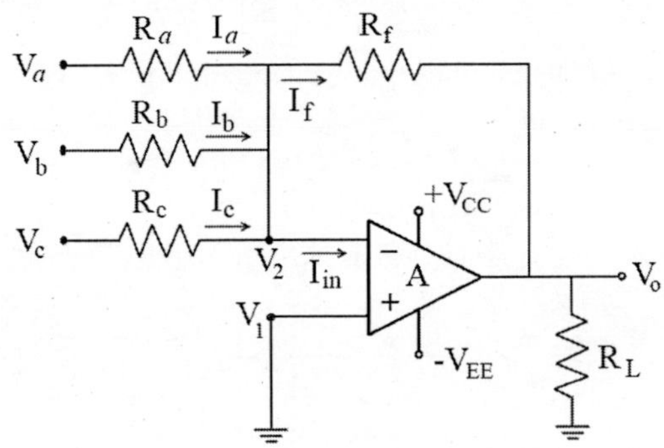

Fig. 4.18: Inverting summing amplifier

By applying KCL at V_2

$$I_a + I_b + I_c = I_f + I_{in}$$

$$I_{in} \cong 0 \quad [\because R_i \cong \infty]$$

$$I_a + I_b + I_c = I_f$$

$$\frac{V_a - V_2}{R_a} + \frac{V_b - V_2}{R_b} + \frac{V_c - V_2}{R_c} = \frac{V_2 - V_o}{R_f}$$

Now $V_2 \cong 0$ $[Using\ virtual\ ground\ concept]$

$$\frac{V_a}{R_a} + \frac{V_b}{R_b} + \frac{V_c}{R_c} = \frac{-V_o}{R_f}$$

$$\therefore V_o = -R_f\left[\frac{V_a}{R_a} + \frac{V_b}{R_b} + \frac{V_c}{R_c}\right]$$

$$V_o = -\left[\frac{R_f}{R_a}V_a + \frac{R_f}{R_b}V_b + \frac{R_f}{R_c}V_c\right]$$

- **For Summing Amplifier:** In a summing amplifier the output voltage is the sum of all the input voltages. The output voltage is given by

$$V_o = -\left[\frac{R_f}{R_a}V_a + \frac{R_f}{R_b}V_b + \frac{R_f}{R_c}V_c\right]$$

If $R_f = R_a = R_b = R_c$, then

$$V_o = -[V_a + V_b + V_c]$$

- **For Scaling Amplifier:** In a scaling amplifier each signal is amplified by a different scale. The output voltage is given by

$$V_o = -\frac{R_f}{R_a}V_a - \frac{R_f}{R_b}V_b - \frac{R_f}{R_c}V_c$$

If $R_a \neq R_b \neq R_c$. Then, it can be used as a scaling amplifier

- **For Averaging Amplifier:** In averaging amplifiers, the output voltage is the average of all the voltages applied to the input terminals. The output voltage V_o is the average of the input voltages.

$$V_o = -\frac{1}{3}(V_a + V_b + V_c)$$

If $R_a = R_b = R_c = R$

$$\therefore V_o = -\frac{R_f}{R}(V_a + V_b + V_c)$$

$$\therefore \quad \frac{R_f}{R} = \frac{1}{n}$$

Where n is the number of inputs.

b) Non-inverting configurations:

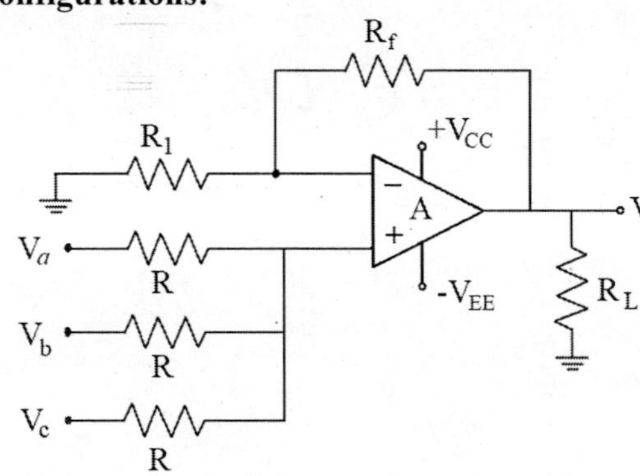

Fig. 4.19: Non-inverting summing amplifier

Applying superposition theorem-

i) Consider V_a acting alone [Ground others]

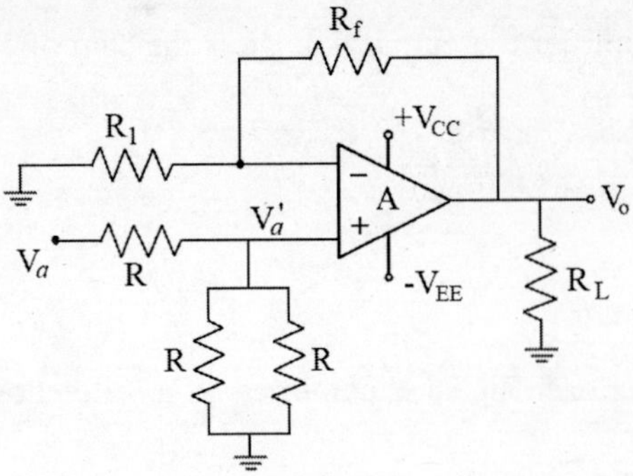

$$V_a' = \frac{V_a \times R/2}{R + R/2} = \frac{V_a}{3}$$

$$V_o' = \left[1 + \frac{R_f}{R_1}\right] V_a'$$

$$V_o' = \frac{V_a}{3}\left[1 + \frac{R_f}{R_1}\right]$$

ii) Considering V_b alone

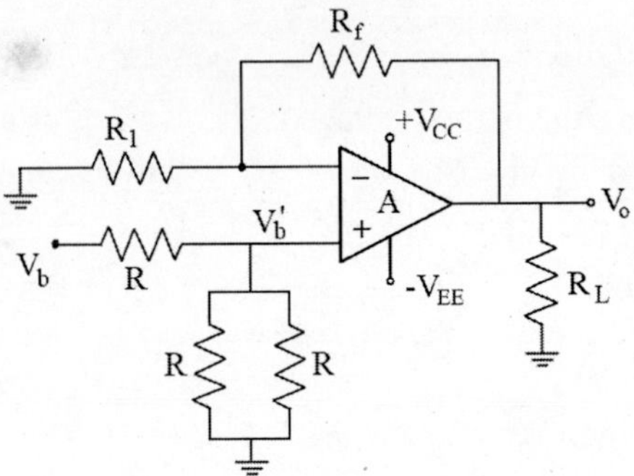

$$V_b' = \frac{V_b}{3}$$

$$V_o'' = \left(1 + \frac{R_f}{R_1}\right)\frac{V_b}{3}$$

iii) Considering V_c alone (Similarly)

$$V'''_o = \left[1 + \frac{R_f}{R_1}\right]\frac{V_c}{3}$$

$$\therefore\ V_o\ (total) = V_o' + V''_o + V'''_o$$

$$= \left[1 + \frac{R_f}{R_1}\right]\left(\frac{V_a + V_b + V_c}{3}\right)$$

- **For summing Amplifier**

Let $R_f = 2R_1$ $\quad\quad \therefore\ \left[1 + \frac{R_f}{R_1}\right] = 3$

$$\therefore\ V_o = V_a + V_b + V_c$$

- **For averaging Amplifier**

Let $\left[1 + \frac{R_f}{R_1}\right] = 1$ $\quad\quad \therefore\ \frac{R_f}{R_1} = 0$

$$\therefore\ R_f = 0$$

$$\therefore\ V_o = \frac{V_a + V_b + V_c}{3}$$

4.6.2. OP-Amp as a Subtractor

Fig. 4.20 shows a subtractor circuit using OP-Amp. Here, an Op-Amp can be used as a subtractor, if all the resistor values are equal. The circuit is analyzed using the superposition principle.

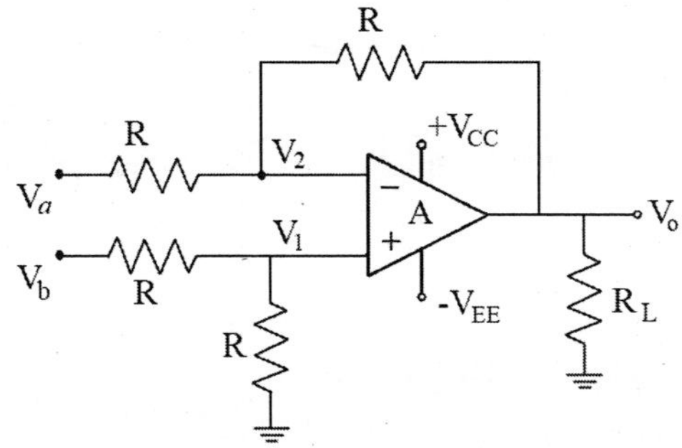

Fig. 4.20: Op-Amp as a Subtractor

a. **Consider V_a acting alone**

Op-Amp works as inverting amplifier.

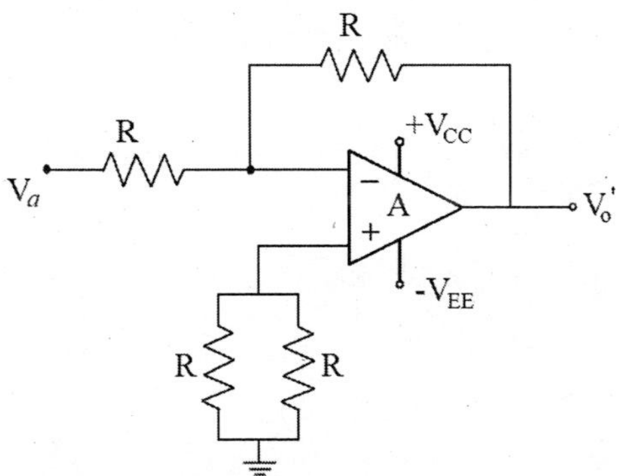

$$V_o' = -\frac{R_f}{R} V_a$$

$$= -\frac{R}{R} V_a$$

$$V_o' = -V_a$$

b. **Consider V_b acting alone**

Op-Amp works as a non-inverting amplifier.

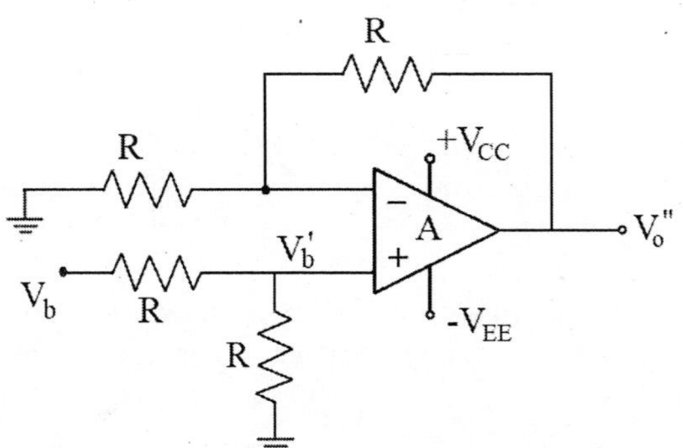

$$V_o'' = \left(1 + \frac{R_f}{R}\right) V_b'$$

$$= \left(1 + \frac{R}{R}\right) \frac{V_b}{2}$$

$$= \frac{2V_b}{2} = V_b$$

$$\therefore V_o = V_o' + V_o''$$

$$V_o = V_b - V_a$$

4.6.3. Op-Amp as an Integrator

Fig. 4.21 shows an integrator circuit using Op-Amp. The circuit can perform the mathematical operation called integration and produces a sine or cosine or ramp of output voltage. This is also called a miller integrator. The output of the integrator is given by:

$$Vo = -\frac{1}{RC}\int_0^t Vin\, dt + Vin\,(0)$$

Where Vin(0) is the integration constant, and RC is called a time constant of an integrator.

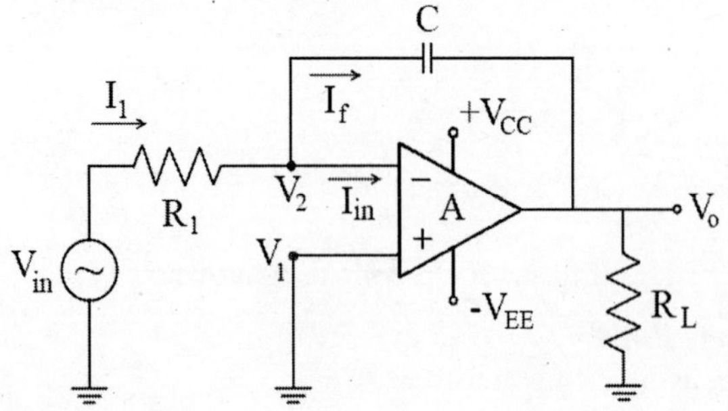

Fig. 4.21: Op-Amp Integrator circuit

By applying KCL at the node V_2,

$$I_1 = I_{in} + I_f$$

$$I_i = I_f \; [\because I_{in} \cong 0 \; as \; R_i = \infty]$$

$$\frac{V_{in} - V_2}{R_1} = C.\frac{d}{dt}(V_2 - V_o) \quad \left[\because \; I_C = C\frac{dV}{dt}\right]$$

Now we have $V_2 = 0$, by the virtual ground concept

$$\therefore \; \frac{V_{in}}{R_1} = C.\frac{d}{dt}(-V_o)$$

By integrating both sides, we get

$$\int \frac{V_{in}}{R_1} dt = C.V_o$$

$$\therefore \; V_o = -\frac{1}{R_1 C}\int V_{in}\,.dt$$

$\therefore$ The gain of the circuit is given by $= -\dfrac{1}{R_1 C}$

4.6.4 Op-Amp as Differentiator

Differentiator is a circuit in which the output voltage is proportional to the instantaneous rate of change of the input voltage. In the differentiator circuit resistor and capacitor is used along with the Op-Amp. Fig.4.22 gives the circuit diagram of a differentiator using Op-Amp. The output of differentiator is given by:

$$Vo = -\,RC\,.\frac{d\,Vin}{dt}$$

Thus, the output voltage of the differentiator is equal to RC time instantaneous rate of change of input voltage to time.

By applying KCL at the node V_2

$$I_C = I_{in} + I_f$$

We know that $I_{in} = 0$ as $R_i = \infty$, so

$$I_C = I_f$$

Current through the capacitor is given by

$$I_C = C\frac{dV}{dt}$$

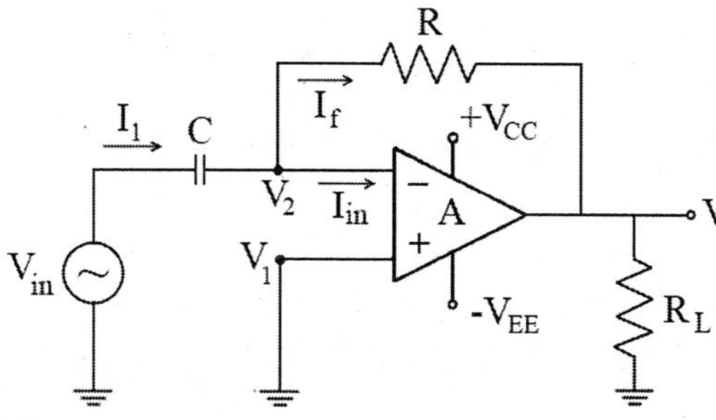

Fig. 4.22: Op-Amp differentiator circuit

$$C\frac{d}{dt}V_{in} = \frac{V_2 - V_o}{R_f}$$

$$C\frac{d}{dt}V_{in} = -\frac{V_o}{R_f} \quad [\because\ V_2 = 0 \text{ By the virtual ground concept}]$$

$$\therefore\ V_o = -R_f.C.\frac{d}{dt}V_{in}$$

If a constantly varying signal such as a sine wave, square wave, or triangular wave to the input of a differentiator amplifier circuit, the final shape of the output signal is dependent upon the RC time constant of the circuit. Fig. 4.23 shows the I/O waveforms of a differentiator circuit using Op-Amp.

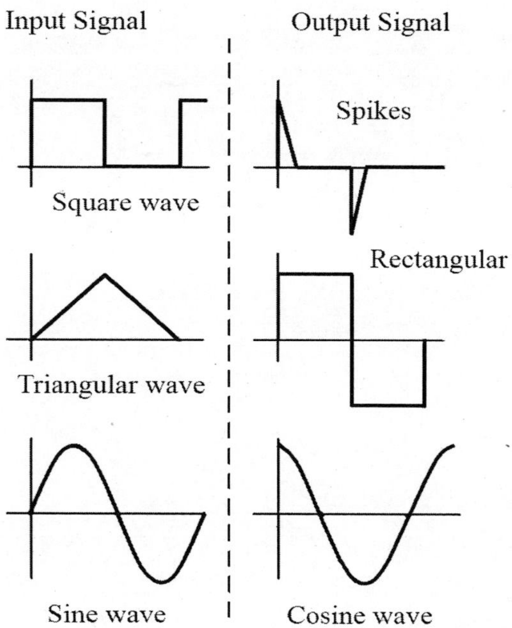

Fig. 4.23: Input and Output waveform of a Differentiator

4.7. Active Filters using Op-Amp

Active filters can be built using Op-Amp and passive components like resistors and capacitors. An active filter circuit additionally uses an amplifier to provide voltage amplification, signal isolation, or buffering of the signals.

A filter that provides a constant output from DC up to a cutoff frequency f_H and then passes no signal above that frequency is called an ideal low-pass filter. The ideal response of a low-pass filter is shown in Fig. 4.24a. A filter that provides or passes signals above a cutoff frequency f_L is a high-pass filter, as shown in Fig. 4.24b. When the filter circuit passes signals that are above one ideal cutoff frequency and below a second cutoff frequency, it is called a bandpass filter, as shown in Fig. 4.24c.

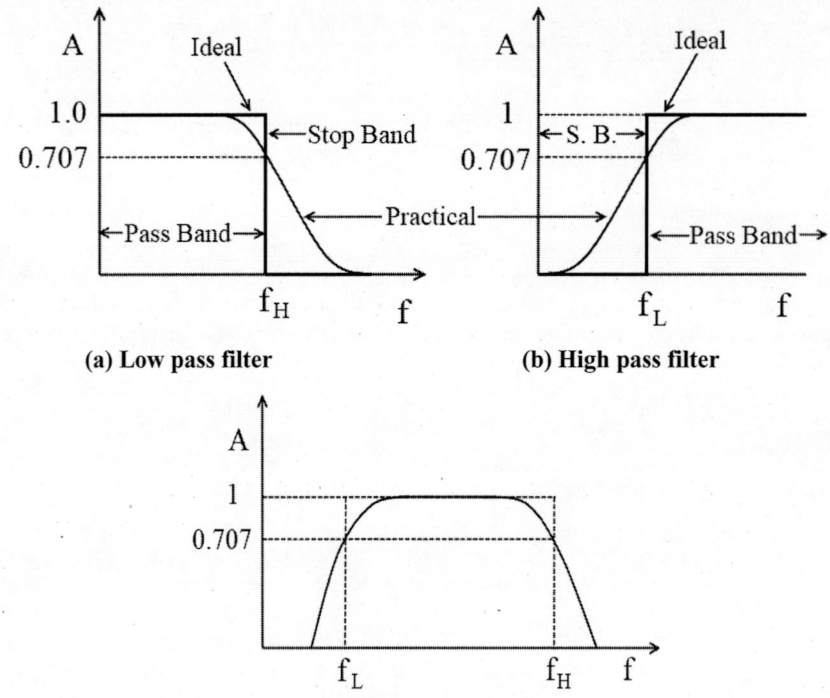

(a) Low pass filter **(b) High pass filter**

(c) Band pass filter

Fig. 4.24: Ideal filter responses

4.7.1. Low pass filter

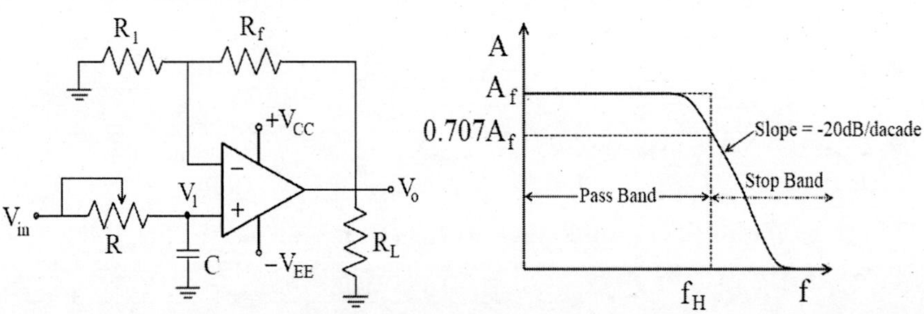

Fig. 4.25: (a) A first-order low pass filter (b) Filter response

A first-order, low-pass filter using a single resistor and capacitor as in Fig. 4.25a has a practical slope of -20 dB per decade, as shown in Fig. 4.25b (rather than the ideal response of Fig. 4.24a). The voltage gain below the cutoff frequency is constant at

$$A_V = 1 + \frac{R_F}{R_1} \tag{4.1}$$

at a constant cut-off frequency of-

$$f_H = \frac{1}{2\pi RC} \tag{4.2}$$

Connecting two sections of a filter as in Fig. 4.26a results in a second-order low-pass filter with a cutoff at -40 dB per decade; closer to the ideal characteristic of Fig. 4.24a. The circuit voltage gain and cutoff frequency are the same for the second-order circuit as for the first-order filter circuit, except that the filter response drops at a faster rate for a second-order filter circuit.

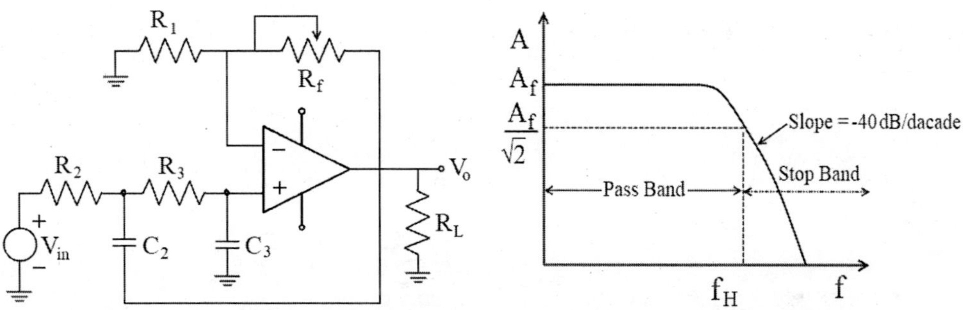

Fig. 4.26 : (a) Second order Low pass filter (b) Filter response

4.7.2. High-pass Active Filter

First- and second-order high-pass active filters can be built as shown in Fig. 4.27a and 4.28a. The amplifier gain is calculated using Eq. (4.1).

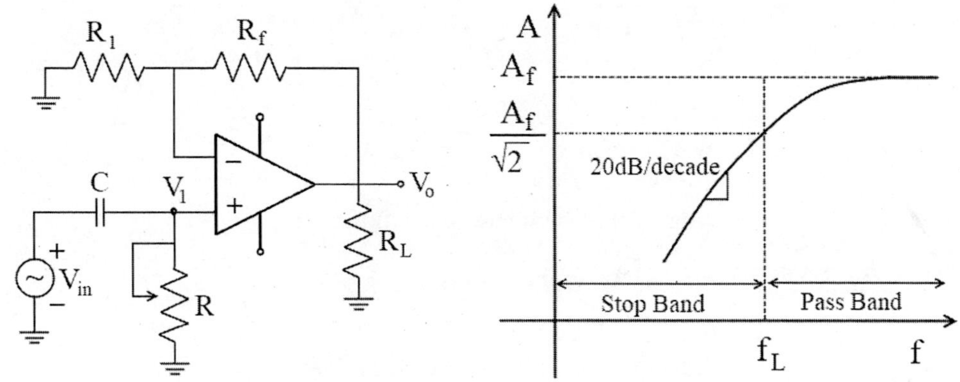

Fig. 4.27 :(a) First order High pass filter (b) Filter response

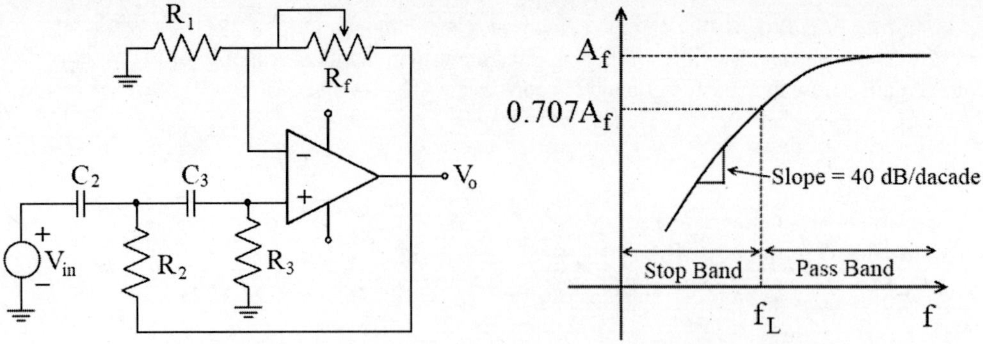

Fig. 4.28 : (a) Second order High pass filter (b) Filter response

The amplifier cutoff frequency is

$$f_L = \frac{1}{2\pi\sqrt{R_2 R_3 C_2 C_3}}$$

4.7.3. Bandpass Filter

Fig. 4.29 shows a bandpass filter using two stages, the first a high-pass filter and the second a low-pass filter, the combined operation being the desired bandpass response.

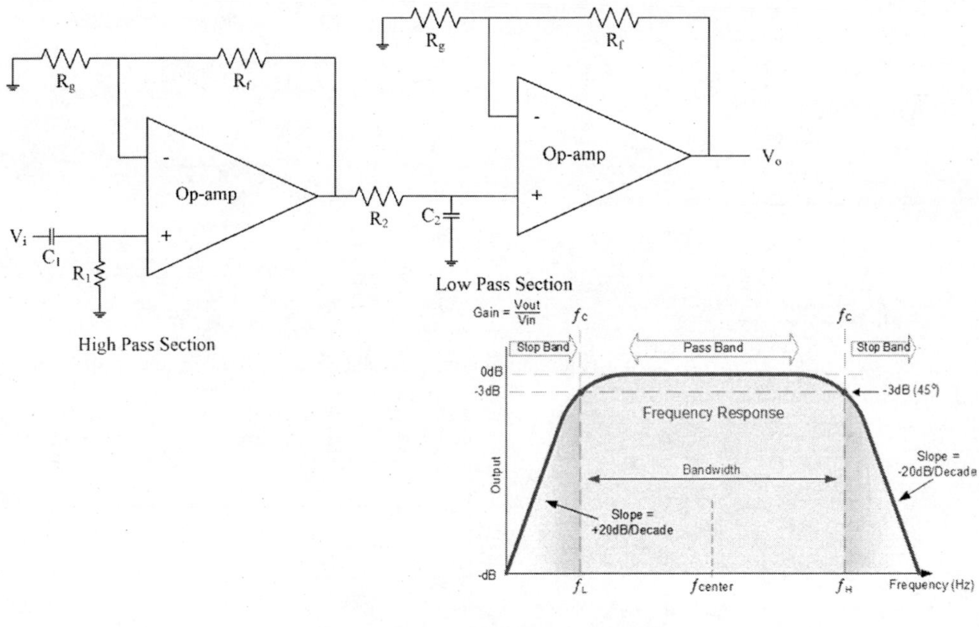

Fig. 4.29: A bandpass active filter

Solved Examples

Ex. 1: Design an inverting amplifier with a closed-loop gain $A_V = -5$. Assume, $V_1 = 0.1\ \sin\omega t$, and a maximum current $i = 5\ \mu A$. The frequency effect can be neglected.

Solution:

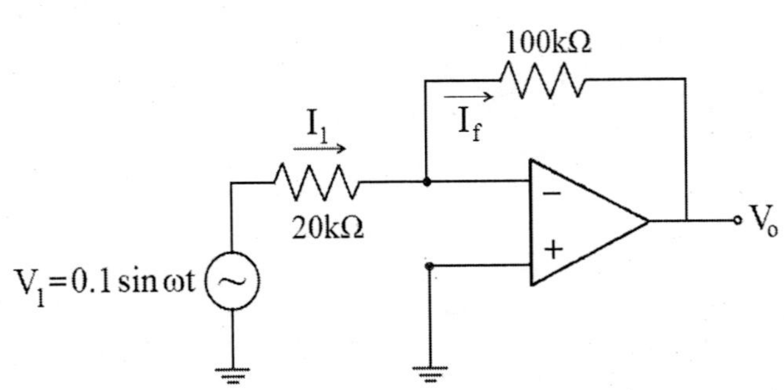

$$R_1 = \frac{V_1}{I_1}$$

$$= \frac{0.1}{5\ \mu A} = 20\ k\Omega$$

Closed loop again

$$A_V = -\frac{R_f}{R_1}$$

$$-5 = -\frac{R_f}{20\ k\Omega}$$

$$R_f = 100\ k\Omega$$

Designed circuit:

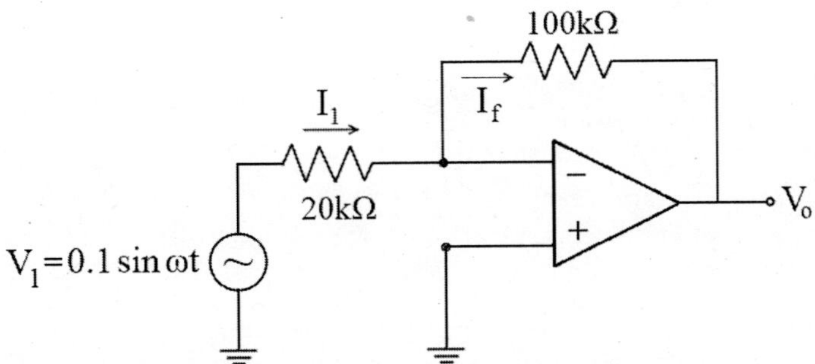

Ex. 2: Calculate the A_{Vf}, R_{if}, R_{of} for the given inverting amplifiers.

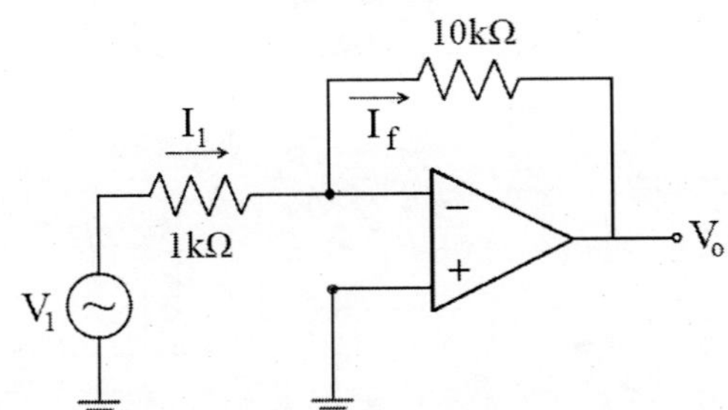

Solution:

i) A_{Vf}

$$A_{Vf} = -\frac{R_f}{R_1} = -\frac{10\ k\Omega}{1\ k\Omega}$$

$$= -10$$

ii) R_{if}

$$R_{if} = R_1 + \left[\frac{R_f}{1 + A_V} \parallel R_i\right]$$

Where:

$$A_V = 2 \times 10^5$$

$$R_i = 2\ M\Omega$$

$$= 1k\Omega + \left[\frac{10\text{k}\Omega}{1 + 2 \times 10^5} \parallel 2\text{M}\Omega\right]$$

$$= 1\ k\Omega + 0.049$$

$$= 1.049\ k\Omega$$

$$R_{if} = 1.049\ k\Omega$$

iii) R_{of}

$$R_{of} = \frac{R_o}{[1 + A_V \beta]}$$

Where:

$$R_O = 75\ \Omega$$

$$A_V = 2 \times 10^5$$

$$\beta = \frac{R_1}{R_1 + R_f}$$

$$= \frac{1\ k\Omega}{1\ \text{k}\Omega + 10\ \text{k}\Omega} = -\frac{1}{11}$$

$$R_{Of} = \frac{75}{1 + (2 \times 10^5)\left(\frac{1}{11}\right)}$$

$$= \frac{75}{18182.8} = 4.12\ m\Omega$$

$$R_{Of} = 4.12\ m\Omega$$

Ex. 3: Find the output voltage for the circuit shown in the figure. If $R = 10\ k\Omega,\ R_1 = 2\ k\Omega$ and $R_2 = 5\ k\Omega$.

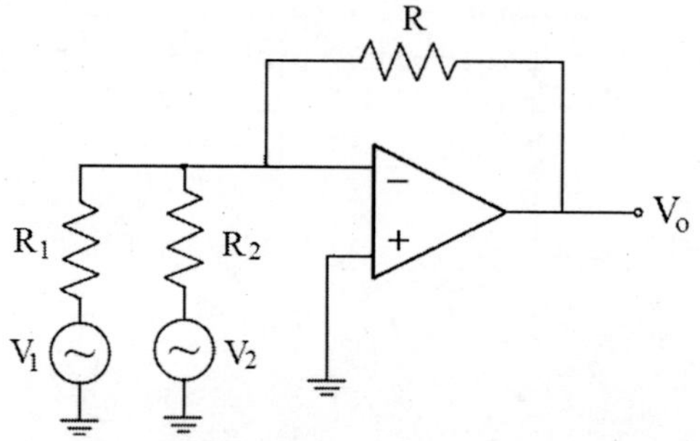

Solution: By applying the superposition theorem, the output voltage V_0

$$V_0 = \left(-\frac{R_f}{R_1}V_1\right) + \left(-\frac{R_f}{R_1}V_2\right)$$

$$V_0 = -5V_1 + (-2V_2)$$

$$V_0 = -[5V_1 + 2V_2]$$

Ex. 4: Design a non-inverting amplifier with a closed-loop gain $A_V = 5$. Assume $V_1 = 0.1 \sin \omega t$, a maximum current $i = 5mA$. The frequency effect is neglected.

Solution:

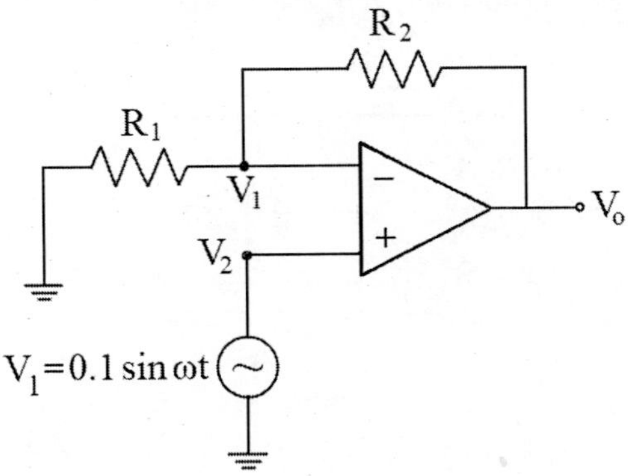

$$V_1 = V_2 \quad \text{\{by Virtual Ground Concept\}}$$

$$R_1 = -\frac{V_i}{I_i} = \frac{+0.1}{+5\ mA}$$

$$R_1 = 20\ k\Omega$$

Closed loop gain A_V

$$A_V = \left(1 + \frac{R_2}{R_1}\right)$$

$$5 = \left(1 + \frac{R_2}{20\ k\Omega}\right)$$

$$R_2 = 4 \times 20\ k\Omega = 80\ k\Omega$$

$$R_2 = 80\ k\Omega$$

Ex. 5: Design an adder using OP-Amp to get output expression as: $V_0 = -[0.2V_1 + V_2 + 20V_3]$, where V_1, V_2 and V_3 are the inputs.

Solution:

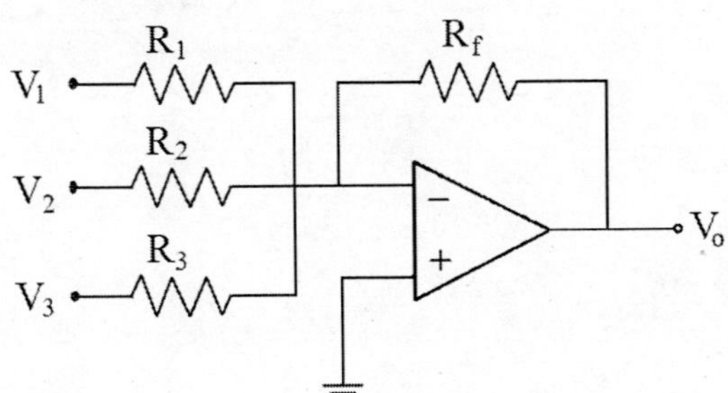

For this inverting adder circuit the expression for output voltage (V_0) is given by

$$V_0 = -\left[\frac{R_f}{R_1}V_1 + \frac{R_f}{R_2}V_2 + \frac{R_f}{R_3}V_3\right]$$

Comparing with the given expression

$$V_0 = -[0.2V_1 + V_2 + 20V_3]$$

And let us assume $R_f = 20k\Omega$, Then

$$R_1 = 100\ k\Omega, \qquad R_2 = 20k\Omega, \qquad R_3 = 1k\Omega$$

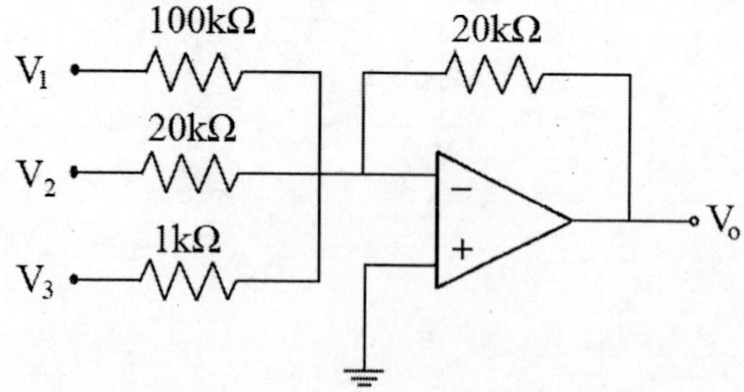

Ex. 6: Find V_0 for the adder–subtractor shown in the figure.

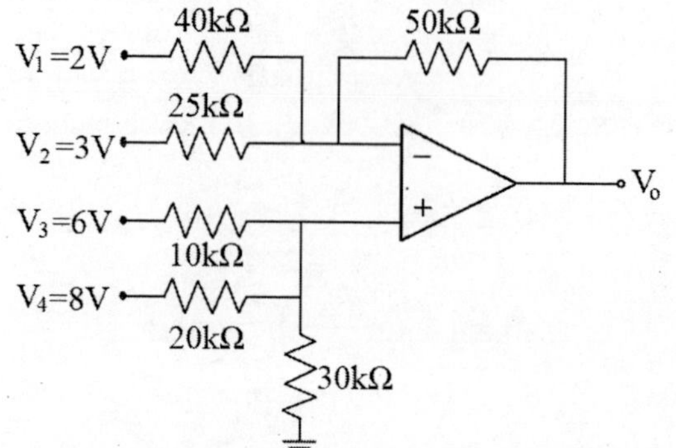

Solution: The sum of inverting configuration is obtained by putting $V_3 = V_4 = 0$

$$V_0' = -V_1\frac{R_f}{R_1} - V_2\frac{R_f}{R_2}$$

$$= -\frac{50}{40}V_1 - \frac{50}{25}V_2$$

$$V_0' = -1.25V_1 - 2V_2$$

Now for a non-inverting adder put $V_1 = V_2 = 0$

By applying the superposition theorem we have,

Considering only V_3 & $V_4 = 0$

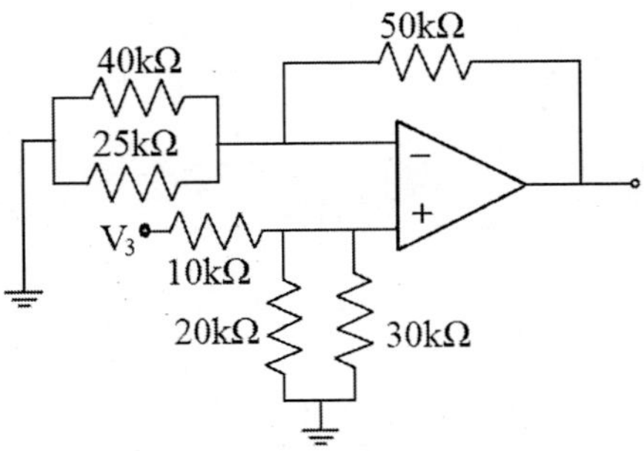

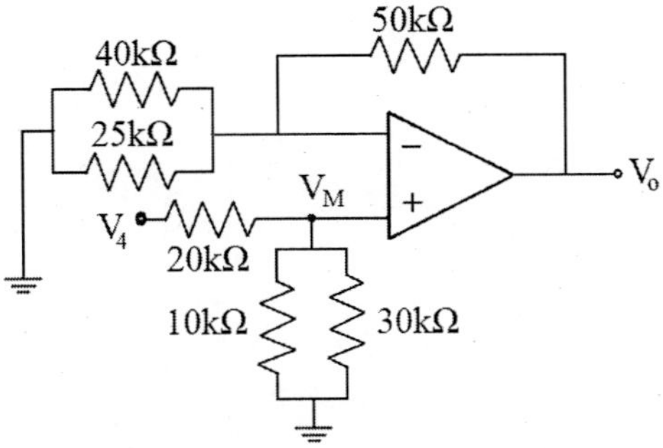

$$V_M = 12\frac{V_3}{10+12} + \frac{7.5V_4}{20+7.5}$$

$$V_M = 0.545V_3 + 0.273V_4$$

The output voltage V_0'' due to V_3 and V_4 is given by

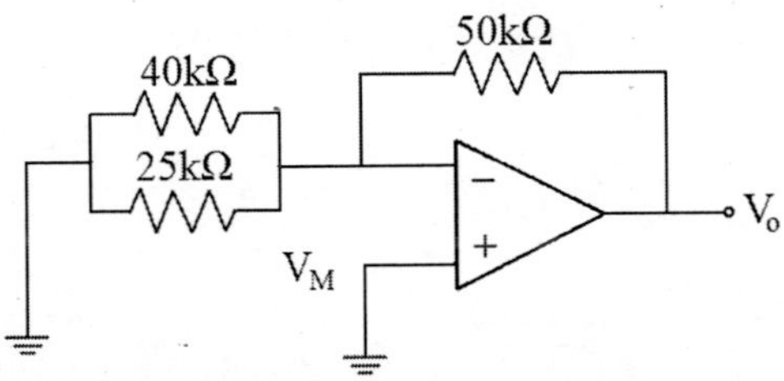

$$V_0'' = \left[1 + \frac{R_f}{R}\right]V_M$$

$$= \left[1 + \frac{50}{15.8}\right][0.545V_3 + 0.273V_4]$$

$$= 2.38V_3 + 1.16V_4$$

Total output voltage

$$V_0 = V_0' + V_0''$$

$$= -1.25V_1 + 2V_2 + 2.38V_3 + 1.16V_4$$

$$= -1.25(2) + (2 \times 3) + (2.38 \times 6) + (1.16 \times 8)$$

$$= 14.7$$

$$V_0 = 14.7\ V$$

Ex. 7: Determine the output for the integrator circuit shown in the figure. For $R_1C_1 = 1$Sec and the input is a step signal.

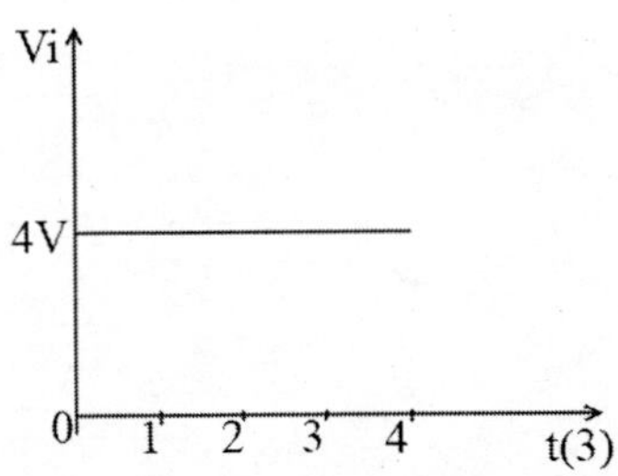

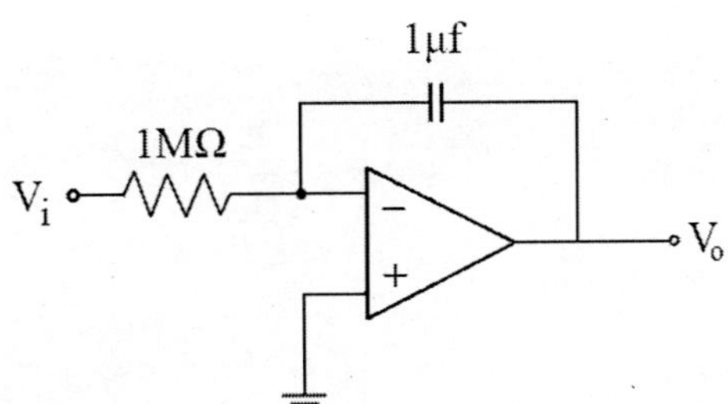

Solution: The input is constant value ranges from $0 < t < 4$, therefore

$$V_0 = -\frac{1}{R_1C_1}\int_0^4 4\,dt$$

$$= \frac{1}{1 \times 10^{-6} \times 1 \times 10^{+6}}\int_0^4 4\,dt$$

$$= -\int_0^4 4\,dt$$

$$= -4[t]_0^4$$

$$= -4[4]$$

$V_0 = -16V$ (Output will go from zero to -16V)

Ex. 8: Design an Op-Amp differentiator that will differentiate an input signal with $f_0 = 200Hz$

Solution:

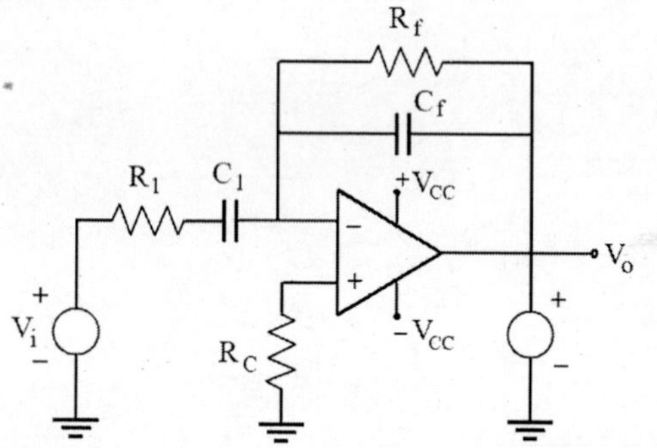

$$fa = f_0 = 200Hz$$

$$fa = \frac{1}{2\pi R_f C_1}$$

Let

$$C_1 = 0.1\mu f, we\ get$$

$$R_f = \frac{1}{2\pi \times 200 \times 0.1 \times 10^{-6}}$$

$$= 795.77\Omega$$

$$R_f = 0.795k\Omega$$

Let

$$R_1 = 0.7k\Omega$$

Since,

$$R_f C_f = R_1 C_1$$

$$C_f = \frac{R_1 C_1}{R_f}$$

$$= \frac{0.7 \times 10^3 \times 0.1 \times 10^{-6}}{0.7 \times 10^3}$$

$$= 0.1 \times 10^{-6}$$

$$C_f = 0.1\ \mu f$$

b) Draw the output waveform for sine wave 2V peak at $200Hz$ applied.

$$V_0 = -R_f C_1 \frac{dV_i}{dt}$$

$$= -[0.7 \times 10^3][0.1 \times 10^{-6}]\frac{d}{dt}(2 \sin 2\pi\, 200t)$$

$$= -[0.7 \times 10^3][0.1 \times 10^{-6}](800\,\pi)(\cos 2\pi\, 200t)$$

$$V_0 = -\,0.175\,(\cos 2\pi\, 200t)\ \text{V}$$

Ex. 9: Design a non-inverting amplifier with specifications of closed-loop gain of 5, limited output voltage of $-5V \le V_0 \le +5V$ and the maximum input current of $50\mu A$.

Solution: The closed-loop gain of the non-inverting amplifier is given by

$$A_V = \frac{V_O}{V_{in}} = \left[1 + \frac{R_2}{R_1}\right] = 5 \ldots\ldots\ldots\ldots\ldots\ldots\ldots (i)$$

Therefore

$$\frac{R_2}{R_1} = 4 \ldots\ldots\ldots\ldots\ldots\ldots\ldots\ldots\ldots\ldots (ii)$$

The output voltage of the amplifier is $V_O = 5V$, by substituting this in equation (i) we get

$$V_{in} = 1V$$

The input current of the non-inverting amplifier is

$$i_{in} = \frac{V_{in}}{R_1} = 50\mu A$$

Then $\quad R_1 = 20k\Omega$

Substituting R_1 in equation (ii), we get R_2 as

$$R_2 = 20k\Omega \times 4 = 80k\Omega$$

From this, we can find the output current i_2 as

$$i_2 = \frac{V_0 - V_{in}}{R_2} = \frac{5V - 1V}{80k\Omega} = 50\ \mu A$$

Thus, the design of a non-inverting amplifier for the given specifications can be drawn as shown in the figure.

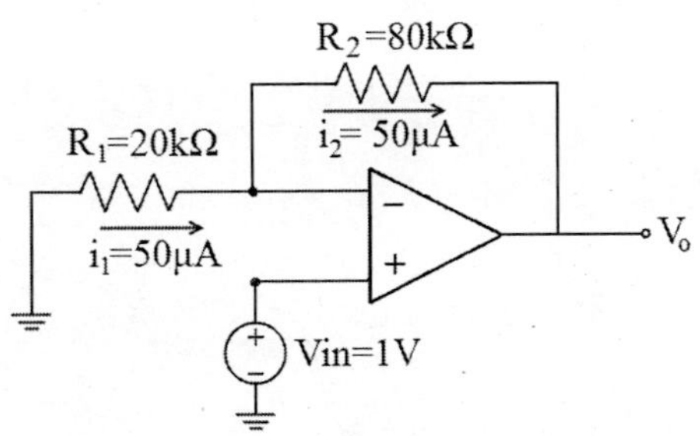

Ex. 10: Consider the difference amplifier shown in the figure. Let $R_1 = R_3 = 10k\Omega$, $R_2 = 20k\Omega$ and $R_4 = 22k\Omega$. Determine V_0 when (i) $V_{I1} = 2V, V_{I2} = -2V$; and

(ii) $V_{I1} = 2V, V_{I2} = 2V$, also determine common mode gain and CMMR.

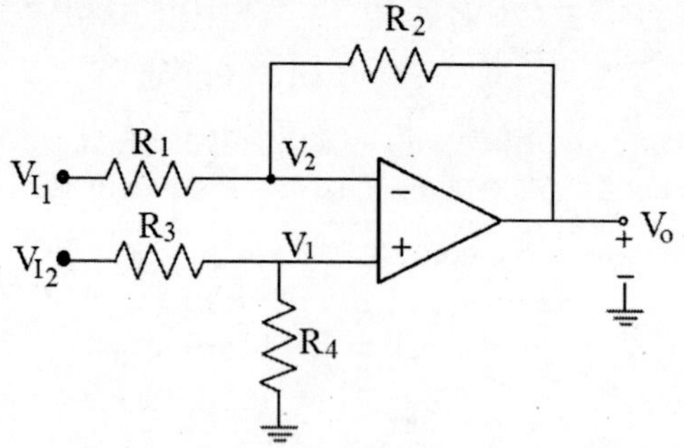

Solution: Given: $R_1 = R_3 = 10k\Omega$

$R_2 = 20k\Omega$

And $R_4 = 22k\Omega$

The output of the difference amplifier is given by

$$V_0 = \left[\frac{R_4}{R_3 + R_4}\right]\left[1 + \frac{R_2}{R_1}\right]V_{I2} - \frac{R_2}{R_1}V_{I1} \dots\dots\dots\dots\dots\dots\dots (1)$$

Case I: $V_{I1} = 2V,\ V_{I2} = -2V$

The output voltage,

$$V_0 = \left[\frac{22}{10+22}\right]\left[1 + \frac{20}{10}\right](-2) - \frac{20}{10}(2)$$

$$V_0 = \left[-\frac{33}{8}\right] - 4 = -8.125V \dots\dots\dots\dots\dots (2)$$

Case II: $V_{I1} = 2V, V_{I2} = 2V$

The output voltage,

$$V_0 = \left[\frac{22}{10+22}\right]\left[1 + \frac{20}{10}\right](2) - \frac{20}{10}(2)$$

$$V_0 = \left[\frac{33}{8}\right] - 4 = 0.125V \dots\dots\dots\dots\dots\dots\dots (3)$$

The common mode input of the difference amplifier is given by

$$V_{CM} = \frac{V_{I2} + V_{I1}}{2} = 2$$

The common mode gain of the difference amplifier is given by

$$A_{CM} = \frac{V_0}{V_{CM}} = 0.062$$

The differential gain of the difference amplifier is given by

$$A_d = \frac{R_2}{R_1} = \frac{20}{10} = 2$$

The common mode rejection ratio of the difference amplifier is given by

$$CMMR = \left|\frac{A_d}{A_{CM}}\right|$$

By substituting the values of differential mode gain and common mode gain values, we get

$$CMMR = \frac{2}{0.062} = 32.25$$

In decibels, it can be expressed as

$$CMMR|_{dB} = 20\log_{10}(CMMR)$$

$$= 20\log_{10}(32.25) = 30.17dB$$

Ex.11: Determine the time constant of the integrator shown in the figure. Consider that the circuit reaches the output voltage of 10V at $t = 2m\ sec$. The step input of $V_I = -1V$ is applied at $t = 0$ sec.

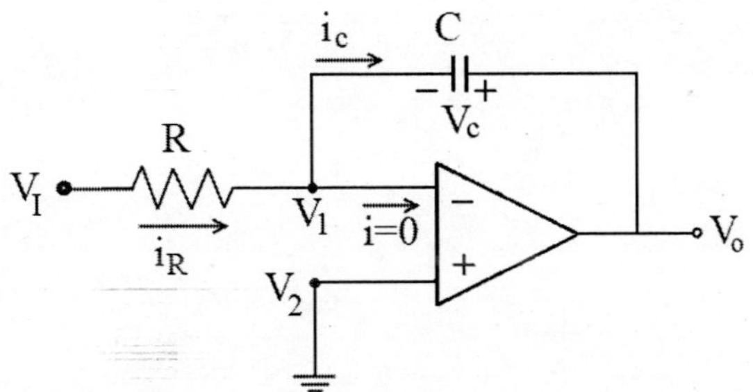

Solution: Given: Output voltage integrator $V_0 = 10V$ at $t = 2m\ sec$

The input Voltage $V_I = (-1)V$ at $t = 0\ sec$

The output voltage of an integrator is defined as

$$V_0 = -\frac{1}{RC}\int_0^t V_I\,dt$$

$$10 = -\frac{1}{RC}\int_0^{2m\ sec} (-1)\,dt$$

$$10 = \frac{1}{RC}[t]_0^{2m\ sec} = \frac{1}{RC}(2m\ sec - 0)$$

$$10 = \frac{2\times10^{-3}}{RC}$$

$$RC = 0.2m\ sec$$

The time constant of the given filter is $RC = 0.2$ m sec.

Ex. 12: Design an integrator circuit that produces the output voltage of 20V at $t = 1m\ sec$ with the step input of $V_I = -1V$ (at $t = 0sec$). Assume that the capacitor voltage is V_C at $t = 0$ sec.

Solution: Given: Output voltage of integrator $V_0 = 20V$ at $t = 1m\ sec$

The input voltage $V_I = (-1)V$ at $t = 0\ sec$

The output voltage of an integrator is defined as

$$V_0 = -\frac{1}{RC}\int_0^t V_I\, dt$$

$$20 = -\frac{1}{RC}\int_0^{1m\ sec} (-1)\, dt$$

$$10 = \frac{1}{RC}[t]_0^{1m\ sec} = \frac{1}{RC}(1m\ sec - 0)$$

$$10 = \frac{10^{-3}}{RC}$$

$$RC = 0.1m\ sec$$

Assume that $R = 1k\Omega$, then,

$$C = \frac{0.1\ m\ sec}{R} = \frac{0.1m\ sec}{1k\Omega}$$

$$C = 0.1\mu F$$

From these values, the design of the integrator can be drawn as shown in the figure below.

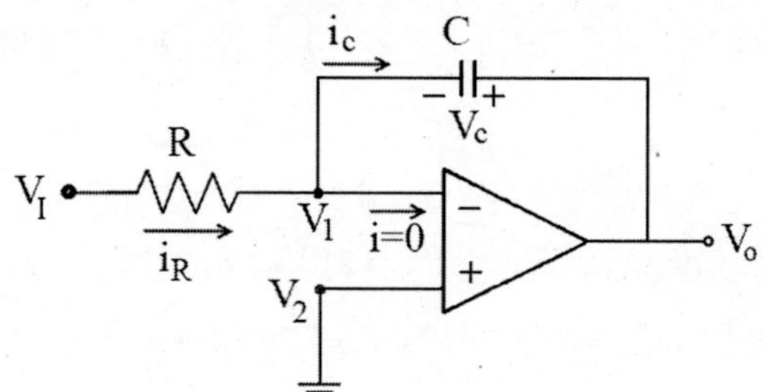

Ex. 11: Design an active low pass filter shown in the figure with a cut-off frequency of 10 KHz.

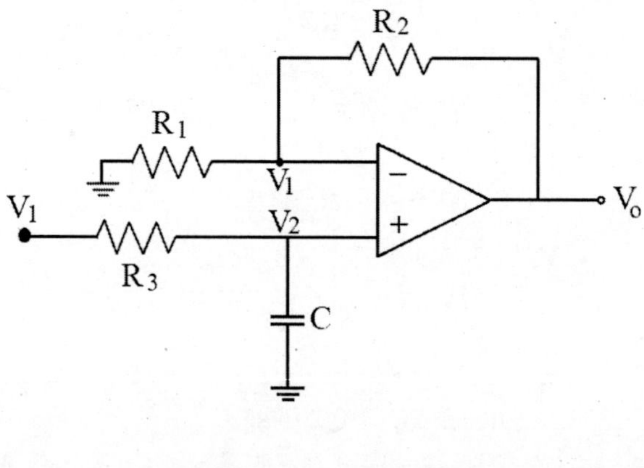

Solution: Given: The cut-off frequency of the low pass filter, $f_C = 10KHz$

The cutoff frequency of the active low-pass filter is defined as

$$f_C = \frac{1}{2\pi R_3 C}$$

Choose the capacitor $C = 0.01\mu F$.

Now the resistance R_3 can be calculated as

$$R_3 = \frac{1}{2\pi f_C C}$$

$$= \frac{1}{2\pi \times (10KHz) \times (0.01\mu F)}$$

$$= 1.59k\Omega$$

The design of an active low-pass filter for a given specification is

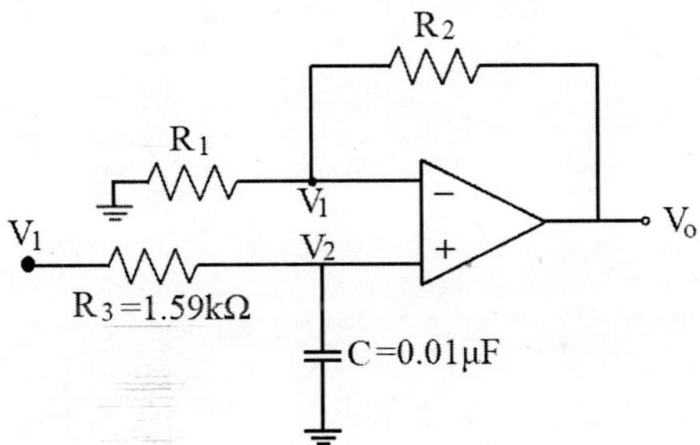

Ex.12. : Design a first-order high pass filter with a cut-off frequency of 2KHz and a voltage gain of 10.

Solution: Given Data: The cut-off frequency, $f_C = 2KHz$

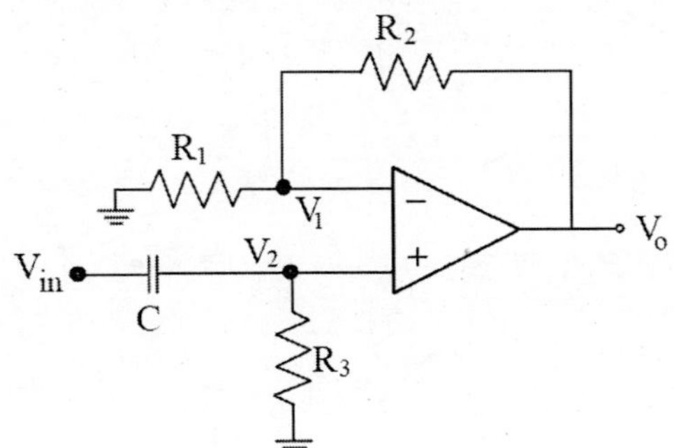

The cut-off frequency of the high-pass filter is

$$f_C = \frac{1}{2\pi R_3 C}$$

If $R_3 = 2k\Omega$, the capacitor C is (Alternately we can assume the value of C and find R_3)

$$C = \frac{1}{2\pi R_3 f_C}$$

$$C = \frac{1}{2\pi \times 2k\Omega \times 2KHz}$$

$$C = 25.13\mu F$$

The voltage gain of the high-pass filter is

$$A_F = \left[1 + \frac{R_2}{R_1}\right]$$

If $R_1 = 10k\Omega$, the feedback resistor R_2 is

$$R_2 = R_1(A_F - 1)$$

$$R_2 = 10k\Omega(10 - 1)$$

$$R_2 = 90k\Omega$$

The design of the first-order high pass filter for the given specifications is shown in the figure.

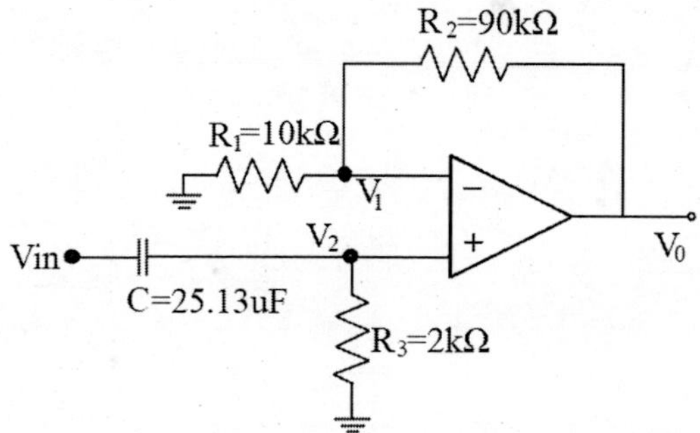

Ex.13: Design an active high pass filter shown in the figure with the cut-off frequency of 20 KHz and the desired passband gain of 15.

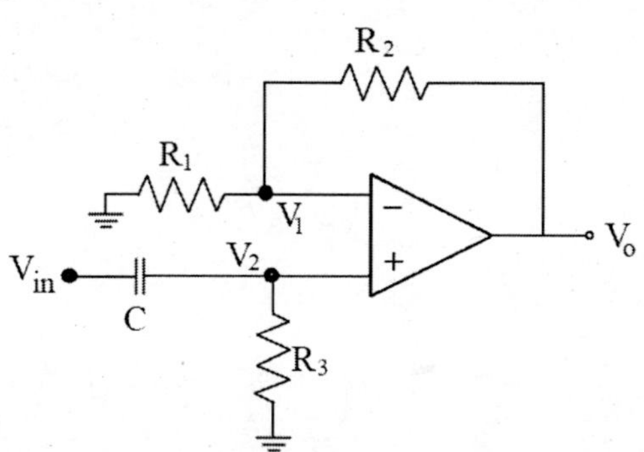

Solution: Given Data: The cut-off frequency high pass filter, $f_C = 20KHz,$

The frequency of the filter, $f = 80kHz$ and

The desired pass band gain $A_F = 15$

The cut-off frequency of the active high-pass filter is defined as

$$f_C = \frac{1}{2\pi R_3 C}$$

Choose the capacitor $C = 0.05\ nf$.

Now the resistance R_3 can be calculated as

$$R_3 = \frac{1}{2\pi f_C C}$$

$$= \frac{1}{2\pi \times (20KHz) \times (0.05\ nf)}$$

$$= 159\ k\Omega$$

The passband gain of the filter is given by

$$A_F = 1 + \frac{R_2}{R_1}$$

Assume that the inverting terminal resistance, $R_1 = 50K\Omega$, then

$$A_F = 1 + \frac{R_2}{R_1}$$

$$15 = 1 + \frac{R_2}{50\ k\Omega}$$

$$R_2 = 700\ k\Omega$$

The design of an active high pass filter for a given specification is shown in the figure below

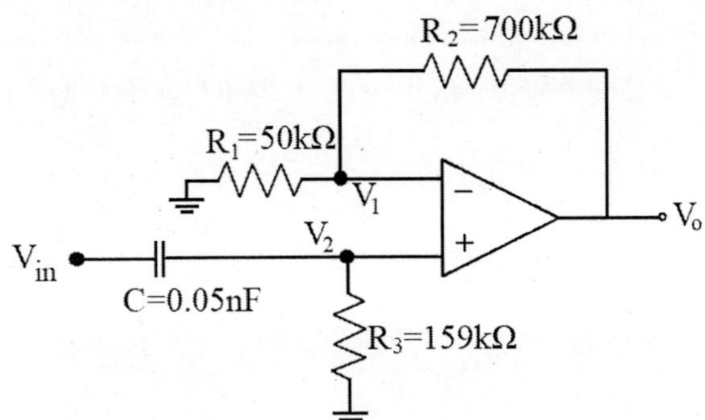

Exercise Questions

1. Draw and explain the block diagram of a typical Op-Amp.
2. Discuss the different characteristics of Op-Amp.
3. Explain the working of the Op-Amp inverting amplifier.
4. Explain the working of the Op-Amp non-inverting amplifier.
5. Draw and explain the circuit of the basic integrator using Op-Amp.

6. Design the Op-Amp circuit which can give the output as $V_O = 2V_1 + 4V_2 - V_3$
7. Design an ideal non-inverting amplifier with a closed-loop voltage gain of $A_V = 10$. When the input voltage is $V_{in} = 0.8V$, the current in any resistor is limited to $100\mu A$.
8. Design a summing amplifier that will produce an output voltage $V_O = -(7\,V_{I1} + 15\,V_{I2} + 10\,V_{I3} + 3\,V_{I4})$. Assume feedback resistance $R_2 = 630k\Omega$.
9. Design an Op-Amp circuit that produces $V_O = V_2 - 3V_1$ with $R_{I1} = R_{I2} = 100k\Omega$.
10. For the difference amplifier shown in the figure determine the common mode rejection ratio if $\frac{R_2}{R_1} = 10$ and $\frac{R_2}{R_4} = 11$.

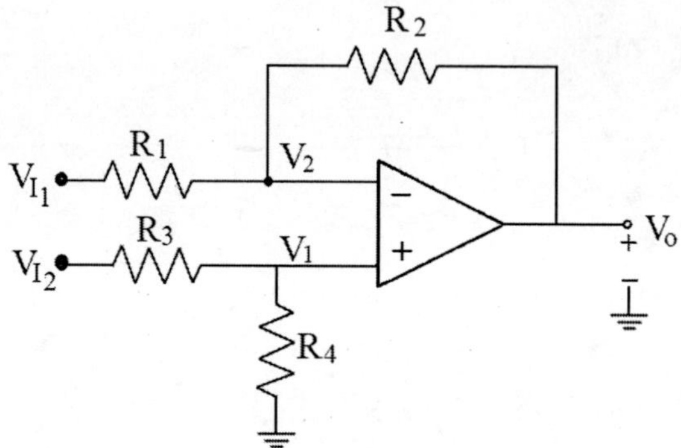

11. For the integrator circuit shown in the figure, determine the output voltage V_O for the input frequency of $10Hz$, $100Hz$, and 1000Hz. Consider that the sinusoidal input peak voltage of the integrator is 1V and the input resistance $R = 150k\Omega$, and the feedback capacitance $C = 1nF$. Assume that the operational amplifier used in the integrator saturates at $\pm 30V$.

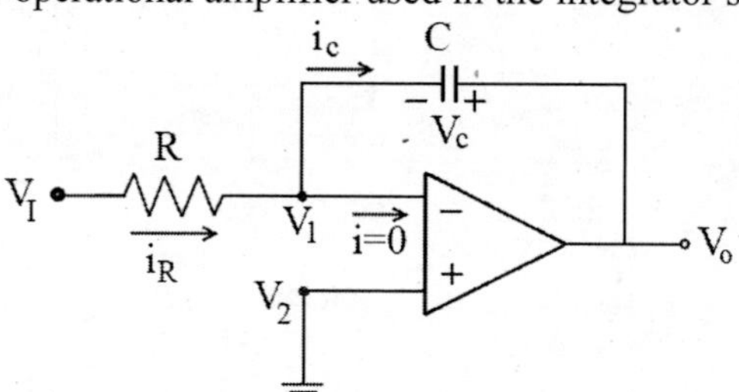

12. Determine the input voltage of an operational amplifier circuit shown in the figure, if the output voltage $V_O = 2V$, feedback resistance $1M\Omega$, and the input resistance $R_1 = 20k\Omega$.

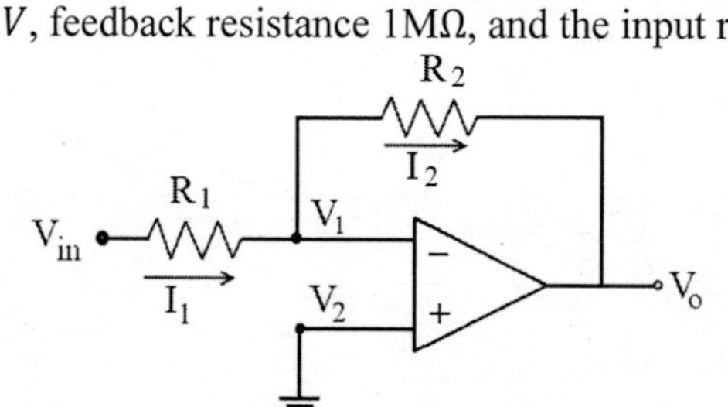

13. Design an active low pass filter shown in fig. with the cut-off frequency of 15 KHz. Calculate the desired passband gain of the filter.

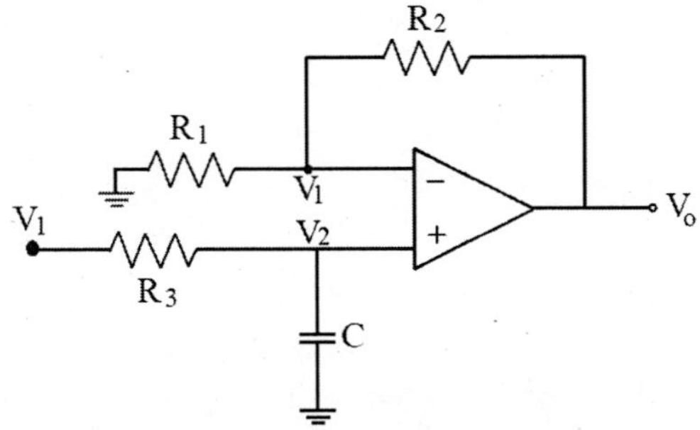

14. Design an active high pass filter shown in the figure with a cut-off frequency of 10KHz.

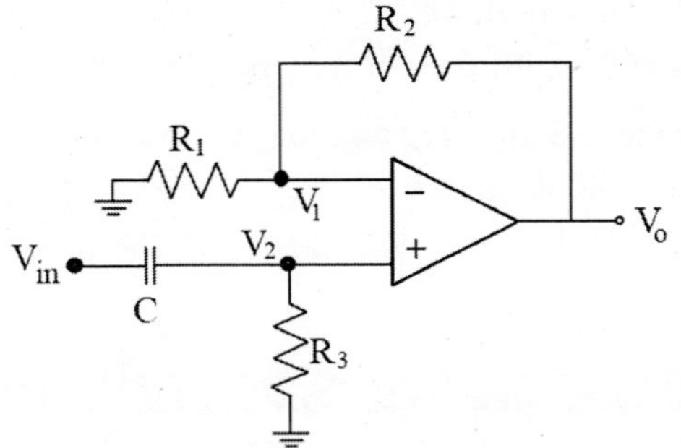

Self-Study Questions

1. Compare active and passive filters.
2. Draw a circuit diagram using Op-Amps to realize $V_O = 5V_1 + 7V_2 - 3V_3$.
3. Discuss the difference in the frequency response of integrator and Low-pass filter, High pass filter, and bandpass filter.
4. Design a summing amplifier to produce the output $V_O = -(3\,V_{Ia} + 12\,V_{Ib} + 15\,V_{Ic} + 18\,V_{Id})$. Assume the feedback resistance $R_2 = 20k\Omega$.
5. List the disadvantages of a basic integrator circuit. How they can be overcome.

Multiple Choice Questions

1. Determine the output from the following circuit

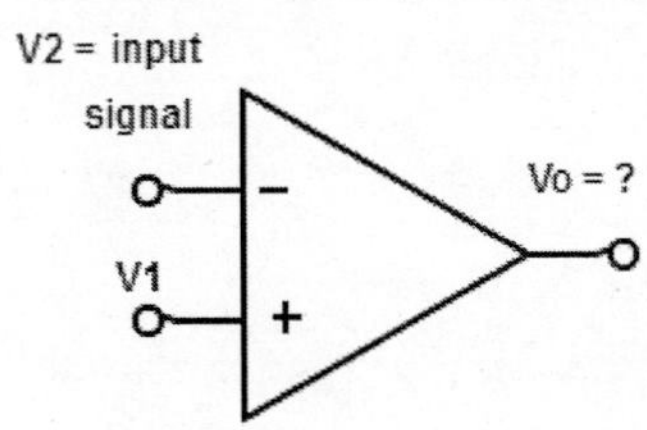

a) 180° in phase with input signal
b) 180° out of phase with input signal
c) Same as that of an input signal
d) Output signal cannot be determined

2. Which of the following electrical characteristics is not exhibited by an ideal Op-Amp?
a) Infinite voltage gain
b) Infinite bandwidth
c) Infinite output resistance
d) Infinite slew rate

3. An ideal Op-Amp requires infinite bandwidth because
a) Signals can be amplified without attenuation
b) Output common-mode noise voltage is zero
c) Output voltage occurs simultaneously with input voltage changes
d) Output can drive an infinite number of device

4. The ideal Op-Amp has infinite voltage gain because
a) To control the output voltage
b) To obtain a finite output voltage
c) To receive zero noise output voltage
d) None of the mentioned

5. Find the output voltage of an ideal Op-Amp. If V_1 and V_2 are the two input voltages
a) $V_O= V_1-V_2$
b) $V_O= A\times(V_1-V_2)$
c) $V_O= A\times(V_1+V_2)$
d) $V_O= V_1\times V_2$

6. How will be the output voltage obtained for an ideal Op-Amp?
a) Amplifies the difference between the two input voltages
b) Amplifies individual voltages input voltages
c) Amplifies products of two input voltage
d) None of the mentioned

7. Which is not the ideal characteristic of an Op-Amp?
a) Input Resistance ≥ 0
b) Output impedance ≥ 0
c) Bandwidth $\geq \infty$
d) Open loop voltage gain $\geq \infty$

8. Which factor determines the output voltage of an Op-Amp?
a) Positive saturation
b) Negative saturation
c) Both positive and negative saturation voltage
d) Supply voltage

9. Op-Amp is a ______ type of amplifier.
   a) Current
   b) Voltage
   c) Power
   d) Resistance

10. Op-Amp is ______ coupled voltage type of amplifier.

a) AC
b) DC
c) ADC
d) DAC

11. The potential output of Op-Amp is _____ times greater than the potential difference of input.

a) 100 times
b) 10000 times
c) 100,000 times
d) 10000000 times

12. Op-Amp is originated from ________ computers.

a) Analog
b) Digital
c) Both a and b
d) None of the above

13. Op-Amp performs which type of mathematical type operations.

a) Linear
b) Non-linear
c) Frequency-dependent
d) All the above

14. Op-Amp was invented by ________.

a) Henry
b) Richard
c) Karl D
d) David

15. In which configuration does the Op-Amp function as a high-gain amplifier?
a) Differential amplifier
b) Inverting amplifier
c) Non-inverting amplifier
d) All of the mentioned

16. How does the open loop Op-Amp configuration classify?
a) Based on the output obtained
b) Based on the input applied
c) Based on the amplification
d) Based on the feedback network

17. What will be the voltage drop across the source resistance of the differential amplifier when connected in an open loop configuration?
a) Zero
b) Infinity
c) One
d) Greater than one

18. The output voltage of an open-loop differential amplifier is equal to
a) Double the difference between the two input voltages
b) Product of voltage gain and individual input voltages
c) Product of voltage gain and the difference between the two input voltages
d) Double the voltage gain and the difference between two input voltages

19. Find the output of the inverting amplifier.
a) $V_o = AV_{in}$
b) $V_o = -AV_{in}$
c) $V_o = -A(V_{in1} - V_{in2})$
d) None of the mentioned

20. What happens if any positive input signal is applied to an open-loop configuration?
a) Output reaches saturation level
b) Output voltage swing's peak to peak
c) Output will be a sine waveform
d) Output will be a non-sinusoidal waveform

21. Why open-loop Op-Amp configurations are not used in linear applications?
a) Output reaches positive saturation
b) Output reaches negative saturation
c) Output switches between positive and negative saturation
d) Output reaches both positive and negative saturation

22. An Op-Amp with negative feedback provides _____ output parameter.

a) Gain
b) Bandwidth
c) Input-output impedance
d) All the above

23. A fully differential amplifier has _______ outputs.

a) Similar
b) Differential
c) Zero
d) None of the above

24. An ideal Op-Amp has _______ input voltage.

a) 1V
b) 3V
c) Grounded
d) Infinite

Answer Keys to MCQ

1	b	7	a	13	c	19	b
2	c	8	c	14	c	20	a

3	**a**	9	**a**	15	**d**	21	**c**
4	**b**	10	**b**	16	**c**	22	c
5	**b**	11	**c**	17	**a**	23	b
6	**a**	12	**a**	18	**c**	24	c

@@@@@@@@@

5. Measuring Instruments & Transducers

RATIONALE

Measuring instruments range from water meters to weighing machines. They are important for trade, consumers, and industry as they ensure accuracy of measurement, transparency, and fairness. Further, as electric flow can't be detected with the naked eye electronic measuring instruments are used to check the flow of electricity. Transducers convert a physical force into an electrical signal so that it can be easily handled and transmitted for measurement.

UNIT OUTCOMES

U5-O1: Unit-5 Learning Outcome-1

To know about the concept of different measuring instruments and their applications.

U5-O2: Unit-5 Learning Outcome-2

To know about the classification and operational characteristics of different types of transducers

LEARNING OBJECTIVES

LO1: To understand the classification and the concept of various measuring instruments.

LO2: To study the D'Arsonval Movement and PMMC-type instruments

LO3: To study the Ohmmeter, Galvanometer, Potentiometers, and Frequency meters.

LO4: To study the characteristics and differences between a sensor and a transducer

LO5: To study different selection criteria of a transducer.

LO5: To study different types of transducers such as Strain gauges, thermistors, LVDT, Hall Effect, and a capacitive transducer.

MAPPING THE UNIT OUTCOMES WITH THE COURSE OUTCOMES

Unit Outcome	EXPECTED MAPPING WITH COURSE OUTCOMES (1- Weak Correlation; 2- Medium correlation; 3- Strong Correlation)						
	CO-1	CO-2	CO-3	CO-4	CO-5	CO-6	CO-7
U5-O1	1	--	--	--	--	3	--
U5-O2	1	--	--	--	--	3	--
U5-O3	1	--	--	--	--	3	--

Interesting Facts:

1. The history of measurement systems in India begins in the early Indus Valley civilization with the earliest surviving samples dated to the 5th millennium BCE. Since early times the adoption of standard weights and measures has been reflected in the country's architectural, folk, and metallurgical artifacts.
2. The person most responsible for bringing sensor technology to the area was Art Zias. Art was a technical writer at Bell Labs while an engineering student in the late fifties. The physics of piezoresistance in silicon and germanium was derived from the work of Phann, Thurston, and Smith at Bell and was chronicled by Art.
3. The first sensor was invented in 1833 by Michael Faraday. It was called the "Faraday's Electromagnetic Sensor" and was used to measure the intensity of an electric current. It worked by measuring the deflection of a compass needle caused by the current flowing through a wire. This invention laid the foundation for the development of other types of sensors such as the Hall Effect sensor which was invented by Edwin Hall in 1879.
4. The hydrophone was invented in 1916. This was the first transducer, which was an electric oscillator that emitted and received a high-frequency signal to indicate the presence of objects in the water.

Video Resources:

Sr	Title	URL	QR Code
1.	Classification of the transducers	https://www.youtube.com/watch?v=VeYMUjmPfy8 https://www.youtube.com/watch?v=QQ9y_T-r370	
2.	Fundamentals of Measurements	https://www.youtube.com/watch?v=pFM9K9JrsU4&list=PLm_MSClsnwm8QtyYpwmQpXjp1XnvuKcRk	

3.	Potentiometers	https://www.youtube.com/watch?v=gApJaU-_ZVI	
4.	Linear Variable Differential Transformers (LVDT)	https://www.youtube.com/watch?v=aqTf195SGrU	
5.	Introduction to Thermistor	https://www.youtube.com/watch?v=zFR385M6ag0	
6.	Introduction to Capacitive Transducers	https://www.youtube.com/watch?v=rrnTdPJpw5A	

The main aim of a measuring system is to present the data understandably to the user/observer. The data presentation device is called an output device or terminating device. There are three types of measuring instruments and they are:

1. Electrical measuring instruments
2. Mechanical measuring instruments.
3. Electronic measuring instruments.

On a broad scale, we can classify these instruments as:

(a) Absolute Measuring Instruments: These instruments provide output in terms of the physical constant of the instruments. Rayleigh's current balance and Tangent galvanometer are absolute instruments.

(b) Secondary Measuring Instruments: These instruments are constructed with the help of absolute instruments. Secondary instruments are calibrated against the absolute instrument.

(c) **Deflection Type Instruments:** In these types of instruments, a pointer of the measuring instrument deflects to measure the quantity under test. The value of the quantity can be estimated by measuring the net deflection of the pointer from its initial position.

(d) **Null Type Instruments:** The null or zero type electrical measuring instruments tend to maintain the position of the pointer stationary by producing an opposing effect.

(e) **Recording Instruments:** These instruments generally use paper to record the output. This type of function is known as the recording function of the instruments.

5.1 D'Arsonval Movement

The D'Arsonval movement is a current sensing mechanism that is used in DC Ammeter, ohm meter, and Voltmeter.

Principle of the D'Arsonval Movement

When an electric current is passed through a coil placed in a magnetic field, it experiences a force. This force causes a torque in the coil that is fixed to a spindle. The spindle can rotate in fixed bearings.

The rotation of the spindle is proportional to the electric current passed through the coil. This torque that is produced is balanced after a movement against the restoring torques of springs. The torque that is produced that tends to rotate the spindle is termed as D'Arsonval Movement.

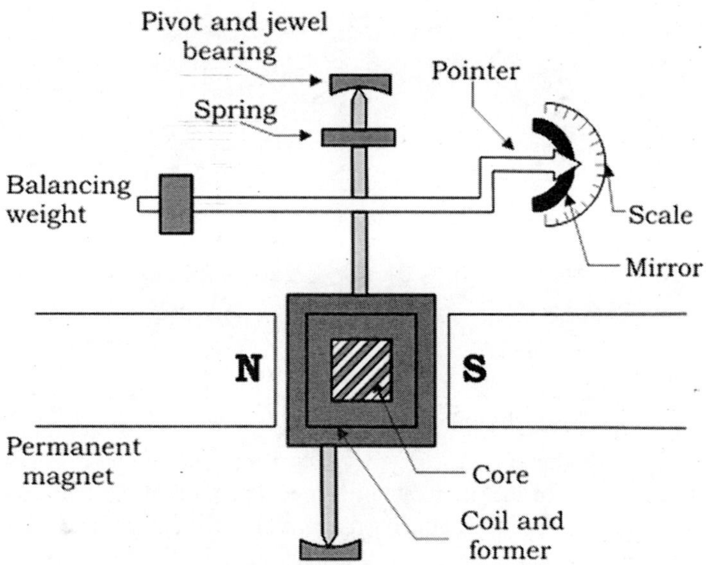

Fig. 5.1: Construction of D'Arsonval Instrument

Description of the D'Arsonval Movement

1. The arrangement consists of a coil that is wound over an iron core (spindle).
2. This spindle is placed between the two poles of a horse shoe magnet.

3. The spindle is attached at its end to the bearings. Spiral and torsional springs are provided for the restoration of the system when the extraction is removed.
4. A pointer is attached to the spindle that can sweep over the calibrated scale.

Operation of D'Arsonval Movement

1. When a current is passed through the coil, it produces a force. Due to this force, a torque is produced in the spindle which rotates it.
2. When the spindle rotates, it moves a pointer making it sweep over the calibrated scale.
3. The spring produces a restoring torque. When this restoring torque becomes equal to the excitation torque, the pointer comes to rest.
4. The rotational movement of the spindle is proportional to the supply (D.C) Current.

5.2 Permanent Magnet Moving Coil Instrument

The instrument which uses the permanent magnet for creating the stationary magnetic field between which the coil moves is known as the permanent magnet moving coil or PMMC instrument. It operates on the principle that the torque is exerted on the moving coil placed in the field of the permanent magnet. The PMMC instrument gives accurate results for DC measurement.

Construction:

Fig 5.2 shows the basic construction of a PMMC-type instrument. The moving coil and permanent magnet are the main part of the PMMC instrument. The parts of the PMMC instruments are explained below in detail.

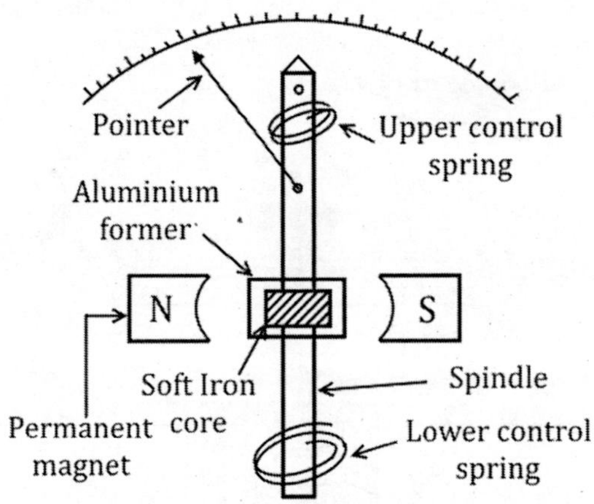

Fig. 5.2: Instrument for Permanent Magnet Moving Coil (PMMC)

Moving Coil – The coil is the current-carrying part of the instruments which is freely moved between the stationary fields of the permanent magnet. The current that passes through the coil deflects it due to which the magnitude of the current or voltage is determined. The coil is mounted on the rectangular former which is made up of aluminum. The former increases the radial and uniform magnetic field between the air gaps of the poles. The coil is wound with the silk cover copper wire between the poles of a magnet.

The coil is mounted on the rectangular former which is made up of aluminum. The former increases the radial and uniform magnetic field between the air gaps of the poles. The coil is wound with the silk cover copper wire between the poles of a magnet.

Magnet System – The PMMC instrument uses a permanent magnet for creating stationary magnets. The Alcomax and Alnico material is used for creating the permanent magnet because this magnet has a high coercive force (The coercive force changes the magnetization property of the magnet). The magnet also has high field intensities.

Control – In the PMMC instrument the controlling torque is because of the springs. The springs are made up of phosphorous bronze and placed between the two jewel bearings. The spring also provides the path for the lead current to flow in and out of the moving coil. The controlling torque is mainly because of the suspension of the ribbon.

Damping – The damping torque is used for keeping the movement of the coil at rest. This damping torque is induced because of the movement of the aluminum core which is moving between the poles of the permanent magnet.

Pointer and Scale – The pointer is linked with the moving coil. The pointer notices the deflection of the coil, and the magnitude of their deviation is shown on the scale. The pointer is made of lightweight material, and hence it is easily deflected with the movement of the coil. Sometimes the parallax error occurs in the instrument which is easily reduced by correctly aligning the blade of the pointer.

Torque Equation:

In moving coil instruments the deflecting torque is given by the expression:

$$T_d = NBldI$$

where N =is number of turns,

B = magnetic flux density in the air gap,

l =the length of the moving coil,

d =the width of the moving coil,

I =the electric current.

Now for a moving coil instrument deflecting torque should be proportional to current, mathematically we can write $T_d = GI$. Thus on comparing we say $G = NBIdl$. At a steady state, both the controlling and deflecting torques are equal. T_c is controlling torque, on equating controlling torque with deflection torque we have $GI = K.x$ where x is deflection thus current is given by

$$I = \frac{K}{G}x$$

Since the deflection is directly proportional to the current therefore we need a uniform scale on the meter for the measurement of the current.

Errors in Permanent Magnet Moving Coil Instruments:

There are three main types of errors:

1. **Errors due to permanent magnets:** Due to temperature effects and aging of the magnets the magnet may lose their magnetism to some extent. The magnets are generally aged by heat and vibration treatment.

2. An error may appear in PMMC Instrument due to the aging of the spring. However, the error caused by the aging of the spring and the errors caused due to the permanent magnet is opposite to each other, hence both errors are compensated with each other.
3. **Change in the resistance of the moving coil with the temperature:** Generally, the temperature coefficient of the value of the coefficient of copper wire in a moving coil is 0.04 per degree Celsius rise in temperature. Due to the lower value of the temperature coefficient, the temperature rises at a faster rate and hence the resistance increases. Due to this significant amount of error is caused.

Advantages:

1. The scale of the PMMC instruments is correctly divided.
2. The power consumption of the devices is very less.
3. The PMMC instruments have high accuracy because of the high torque-weight ratio.
4. The single device measures the different ranges of voltage and current. This can be done by the use of multipliers and shunts.
5. The PMMC instruments use a shelf shielding magnet which is useful for aerospace applications.

Disadvantages:

1. The PMMC instruments are only used for the direct current. The alternating current varies with time. The rapid variation of the current varies the torque of the coil. But the pointer cannot follow the fast reversal and the deflection of the torque. Thus, it cannot use for AC.
2. The cost of the PMMC instruments is much higher as compared to the moving coil instruments.

5.3 Ohmmeter

The meter which measures the resistance and the continuity of the electrical circuit and its components such type of meter is known as the ohmmeter. It measures the resistance in ohms. The micro-ohmmeter is used for measuring the low resistance and the mega-ohmmeter measures the high resistance of the circuit. The ohmmeter is very convenient to use but less accurate. There are many types of ohmmeters available such as

1. Series ohmmeter.
2. Shunt ohmmeter.
3. Multi-range ohmmeter.

Working Principle:

The instrument is connected to a battery, a series of adjustable resistors, and an instrument that gives the reading. The resistance to be measured is connected at terminal ab. When the circuit is completed by connecting output resistance, the circuit current flows and so the deflection is measured.

When the resistance to be measured is very high then the current in the circuit will be very small and the reading of that instrument is assumed to be the maximum resistance to be measured.

When the resistance to be measured is zero then the instrument reading is set to zero position which gives zero resistance.

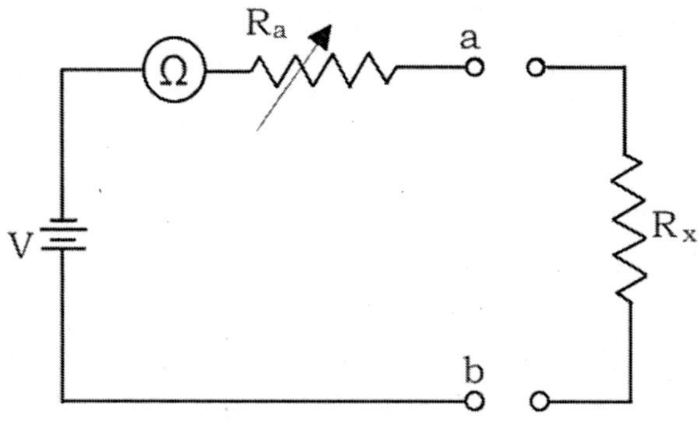

Fig. 5.3: Ohmmeter

5.3.1 Series Ohmmeter

In a series ohmmeter, the measuring resistance component or circuit is connected in series with the meter. The value of resistance is measured through the D'Arsonval movement connected in parallel with the shunt resistor R_2. The parallel resistance R_2 is connected in series with the resistance R1 and the battery. The component whose resistance is used to be measured is connected in series with the terminal A and B. The circuit diagram of the series ohmmeter is shown in Fig. 5.4.

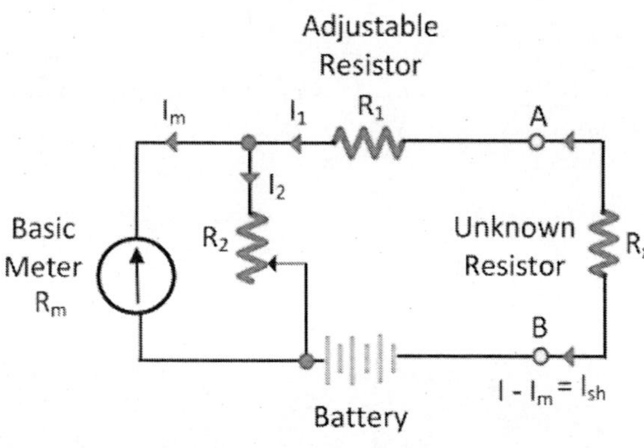

Fig. 5.4:A Series ohmmeter

When the value of unknown resistance is zero the large current flow through the meter. In this condition, the shunt resistance is adjusted until the meter indicates the full load current. For full load current, the pointer deflects towards zero ohms. When the unknown resistance R_x has been removed from the circuit the resistance of the circuit becomes infinite and no current flow through the circuit.

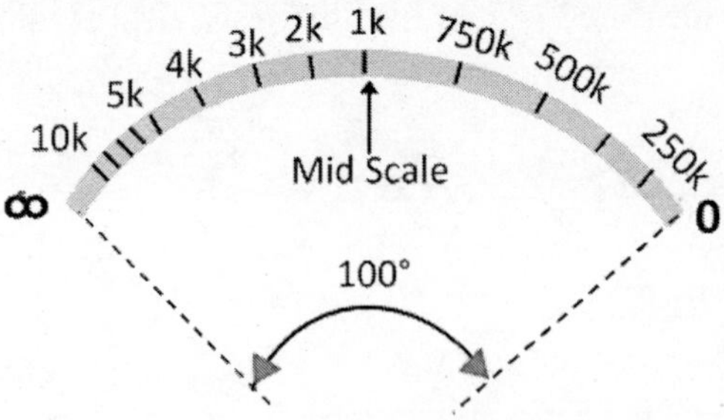

Fig. 5.5: Scale of a series ohmmeter

The pointer of the meter deflects towards the ∞ (infinity). The meter shows the infinite resistance at zero current and the zero resistance when full range current flows through it. When the unknown resistance is connected in series with the circuit and if their resistance is high, then the pointer of the meter deflects toward the left. If the resistance is low, then the pointer deflects toward the right.

5.3.2 Shunt Ohmmeter

The meter in which the measuring resistance is connected in parallel with the battery is known as the shunt ohmmeter. It is mainly used for measuring low-value resistance. The circuit diagram of the shunt ohmmeter is shown in Fig. 5.6.The battery (E), basic meter (R_m), and adjustable resistance are the main components of the shunt ohmmeter. The unknown resistance is connected across terminals A and B. When the value of unknown resistance is zero the meter current becomes zero. And if the resistance becomes infinite (i.e., the terminals A and B are open) then the current passes through the battery and the pointer shows the full-scale deflection toward the left. The shunt-type ohmmeter has a zero mark (no current) on the left of the scale and an infinity mark on the right side.

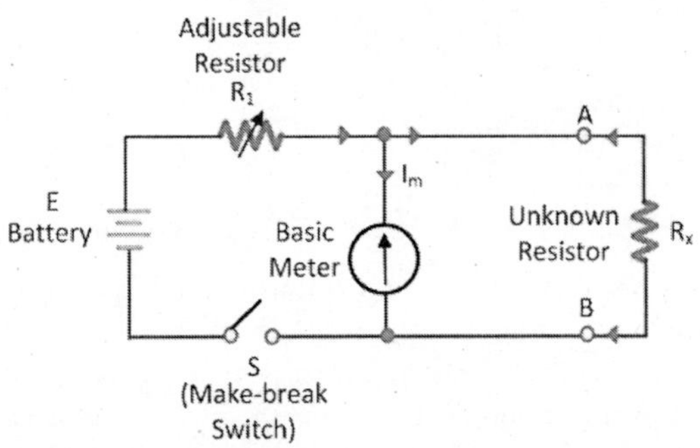

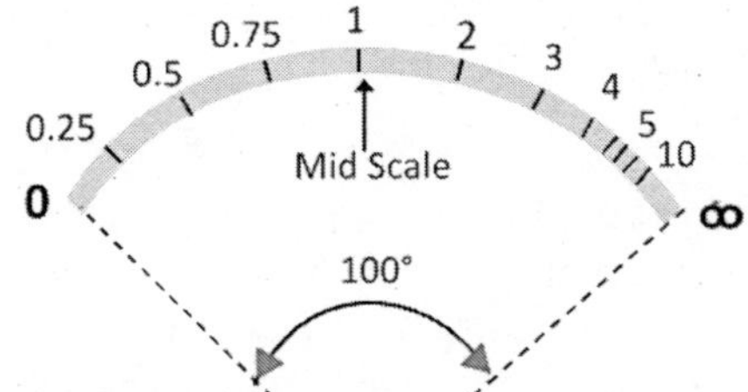

Fig. 5.6:(a) Shunt type ohmmeter **(b) Scale of a shunt ohmmeter**

5.3.3Multi range Ohmmeter

The range of this type of ohmmeter is very high. The meter has an adjuster which selects the range according to need.

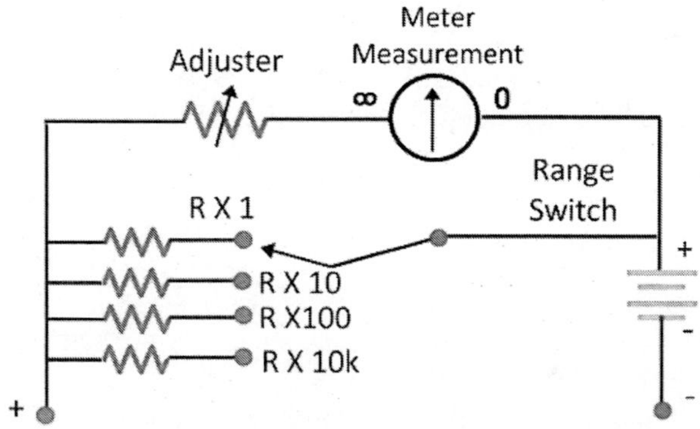

Fig. 5.7: Multirange Ohmmeter

For example, consider we use the meter for measuring the resistance under 10 ohms. For this first, we have to set the range of 10 ohms. The resistance whose value is used to be measured is connected in parallel with the meter. The magnitude of the resistance is determined through the deflection of the pointer.

5.4 Galvanometer

An electromechanical instrument that is used for noticing and signifying an electric current is known as a galvanometer It works as an actuator by generating a rotational deflection in reply to the flow of current throughout a coil in a stable magnetic field. The construction of the galvanometer is shown below. The main parts of this instrument mainly include the suspension, moving coil, and stable magnet.

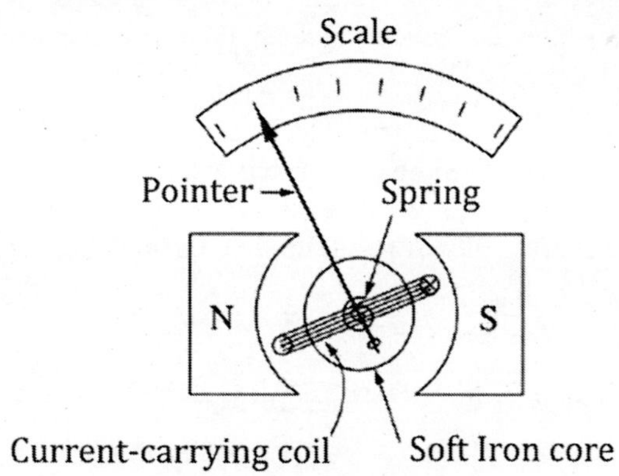

Fig. 5.8: Galvanometer

- ***Moving Coil:*** This is the current-carrying element in the galvanometer. This coil is in a circular otherwise rectangular shape with the no. of twists of copper wire. This coil moves freely between the stable magnet poles. The iron core gives the low reluctance flux lane & therefore gives the tough magnetic field for the twist to move in.
- ***Suspension:*** The balancing of this coil can be done through a plane ribbon. This ribbon supplies the flow of current toward the coil. The other coil which carries the current is the lower suspension and the torque effect of this can be negligible. The designing of the upper suspension coil can be done with a gold wire or copper wire in the ribbon form.

However, the strength of this wire is not extremely tough so the galvanometer handles cautiously without any pulls.

- ***Mirror:*** The suspension in the galvanometer includes a little mirror that throws the ray of light, which is located on the scale where the deflection can be measured.
- ***Torsion Head:*** This is used to control the location of the coil as well as to adjust the setting of zero settings.

Working Principle of Galvanometer

The main function of the galvanometer is to decide the existence, direction, as well as electric current strength in a conductor. This works on the rule of converting energy from electrical to mechanical. Once the current is supplied in a magnetic field, a magnetic torque can be experienced. If it is open to turning below a controlling torque, then it turns by an angle that is proportional to the flow of current through it. This instrument is a kind of ammeter, used to detect and measure electric current.

Whenever a galvanometer is allied to a circuit, then the flow of current will be there in the coil. As the coil is delayed within a magnetic field, then a deflecting torque works upon it. Because of this torque, a coil in the galvanometer will start revolving from its place.

When the coil spins, the springs for controlling will be twisted & a stretchy restoring torque can be developed within them, after that it resists the revolving of the coil.

The coil's rotation angle will be proportional to the torque. As the restoring torque turns into equivalent to the deflecting torque, then the coil relaxes in a stable position. A galvanometer is mainly used in various electrical circuits for detecting current as well as in experiments to decide the null point.

If a heavy current flows through the coil in the galvanometer, then the pointer in this may strike the stop pin because of a very large deflection. So the coil in the galvanometer may blaze because of the extreme heat which is generated.

Hence, this can be protected from these possible harms by using a wide wire otherwise by connecting a copper strip in parallel with its coil which is known as a shunt. When compared to the coil's resistance, its resistance is extremely small. Thus, most of the current flow is supplied through the shunt and some amount of current is supplied through the coil. Therefore, there is no chance of harm to the coil.

Advantages and Disadvantages

The advantages of the galvanometer include the following-

- They will not affect by a strong magnetic field
- Accurate & reliable
- Scales of this are uniform

The disadvantages of the galvanometer include the following-

- Overload can spoil any kind of galvanometer.
- The temperature change will cause a change in restoring torque.
- We cannot change the restoring torque easily.
- These cannot be used for AC quantities measurement.

Applications

- It is used to detect the flow of the current's direction within the circuit and also determines the null point.
- To determine the voltage between two points.
- It can be used in control systems, laser engraving, laser TVs, laser sintering, laser displays, etc.
- It is used in CD/DVD players & hard drives for controlling the position of head servos.
- They are used in a film camera to get the readings of photoresistors in the metering mechanisms

5.5 Potentiometer

A potentiometer (also known as a pot or potmeter) is defined as a 3-terminal variable resistor in which the resistance is manually varied to control the flow of electric current. A potentiometer acts as an adjustable voltage divider.

Working:

A potentiometer is a passive electronic component. Potentiometers work by varying the position of a sliding contact across a uniform resistance. In a potentiometer, the entire input voltage is applied across the whole length of the resistor, and the output voltage is the voltage drop between the fixed and sliding contact as shown in Fig. 5.9.

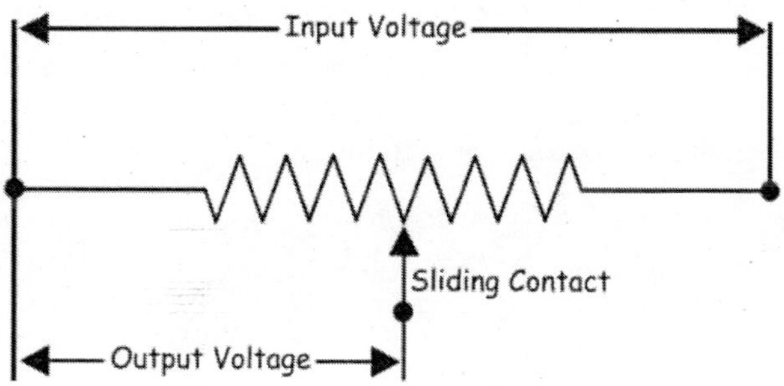

Fig.5.9: Concept of a potentiometer

A potentiometer has two terminals of the input source fixed to the end of the resistor. To adjust the output voltage the sliding contact gets moved along the resistor on the output side. This is different from a rheostat, where here one end is fixed and the sliding terminal is connected to the circuit. The basic working principle of a potentiometer is quite simple. Suppose we have connected two batteries in parallel through a galvanometer. The negative battery terminals are connected and positive battery terminals are also connected through a galvanometer as shown in Fig. 5.10.Here, if the electric potential of both battery cells is the same, there is no circulating current in the circuit and hence the galvanometer shows null deflection. The working principle of the potentiometer depends upon this phenomenon.

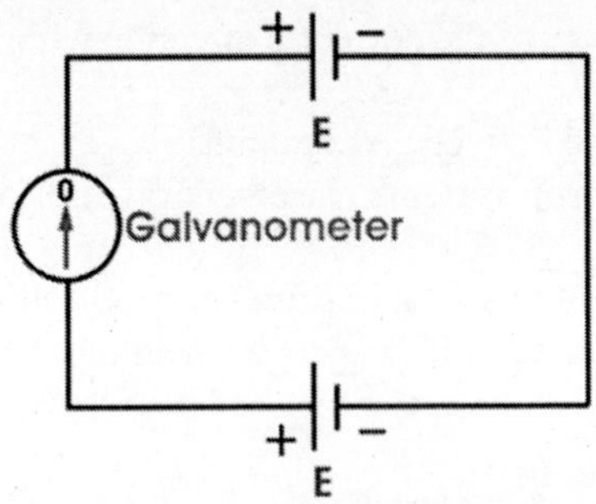

Fig. 5.10: Potentiometer

Types of Potentiometer:

The followings are the main types of potentiometers:

Rotary Potentiometers:

The rotary type potentiometers are used mainly for obtaining adjustable supply voltage to a part of electronic and electrical circuits. This type of potentiometer has two terminal contacts between which a uniform resistance is placed in a semi-circular pattern. The device also has a middle terminal which is connected to the resistance through a sliding contact attached to a rotary knob. By rotating the knob one can move the sliding contact on the semi-circular resistance. The voltage is taken between a resistance end contact and the sliding contact. Fig. 5.11 shows a rotary type of potentiometer.

It is used in substation battery chargers to adjust the charging voltage of a battery. The volume controller of a radio transistor is a popular example of a rotary potentiometer where the rotary knob of the potentiometer controls the supply to the amplifier.

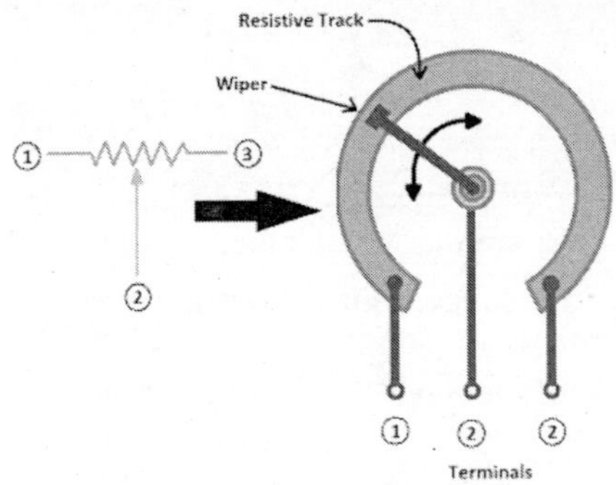

Fig. 5.11: A rotary potentiometer

Linear Potentiometers:

In the linear potentiometers, instead of rotary movement, the sliding contact gets moved on the resistor linearly. Here two ends of a straight resistor are connected across the source voltage. A sliding contact can slide on the resistor through a track attached to the resistor. The terminal connected to the sliding is connected to one end of the output circuit and one of the terminals of the resistor is connected to the other end of the output circuit.

This type of potentiometer is mainly used to measure the voltage across a branch of a circuit, for measuring the internal resistance of a battery cell, for comparing a battery cell with a standard cell, and in our daily life; it is commonly used in the equalizer of music and sound mixing systems.

5.6. Frequency Meters

Frequency meters are the indicating instruments that measure the frequency of electrical energy. This electrical energy may be AC or DC or in a form of various signals or waves produced by various circuits. The working principle of the moving coil frequency meter is the variations in an electric current drawn by inductive and non-inductive circuits are connected in parallel. The different types of frequency meters are:

- Moving Iron Frequency Meter
- Electrodynamic Frequency Meter
- Vibrating reed Frequency Meter

5.6.1. Moving Iron Frequency Meter

Moving iron frequency meters are the meters in which the two coils are fixed and a moving iron is attached to the spindle. This meter depends on the variations in an electric current drawn by inductive and non-inductive circuits connected in parallel. The current flows from these circuits when the frequency changes its value.

Construction:
This meter consists of two fixed coils A and B such that their magnetic axes are perpendicular to each other. A long and soft iron needle in pivoted at their centers. This circuit remains balanced at the supply frequency to be measured. Coil A consists of a series resistance R_A and a reactance L_A in parallel and coil B consists of a series resistance R_B and a reactance L_B in parallel. The series inductance helps to suppress higher harmonics in the current waveform which helps to minimize the waveform errors in the indication of the instrūments.

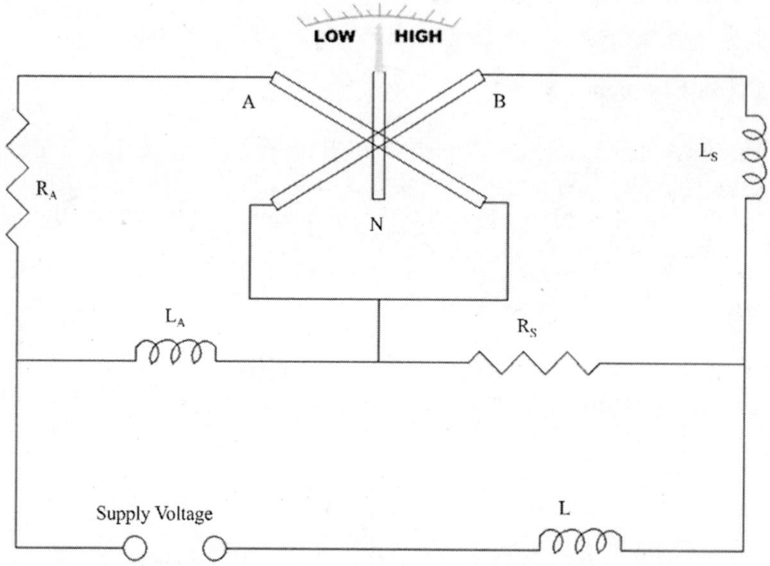

Fig. 5.12: Construction of a moving iron frequency meter

Working: When the supply is connected to the meter, the current pass through coils A and B and these two coils produce opposing torques. When the supply frequency increases then the current of coil A increases and decreases in coil B. The iron needle lies more nearly to the magnetic axis of coil A. For low frequencies, the current of coil B increases and the current of coil A decreases.

5.6.2. Electrodynamic Frequency Meter

Electrodynamic frequency meters/moving coil frequency meters are radiometer types of instruments. These meters are used to measure the frequency of high voltage ranges and too low voltage ranges. These frequency meters provide accurate frequency readings. These frequency meters consist of two moving coils and a rectifier circuit.

Construction:
This frequency meter consists of two moving coils connected at right angles on a shaft and a pointer is also connected to this shaft. These two moving coils are connected with their bridge rectifiers. This circuit consists of a capacitor which is connected in series with the bridge rectifier of the first moving coil C_1 connected to the DC Supply. The direct current (rectified current) flows through a series of resistance to the bridge rectifier and to the second coil C_2.

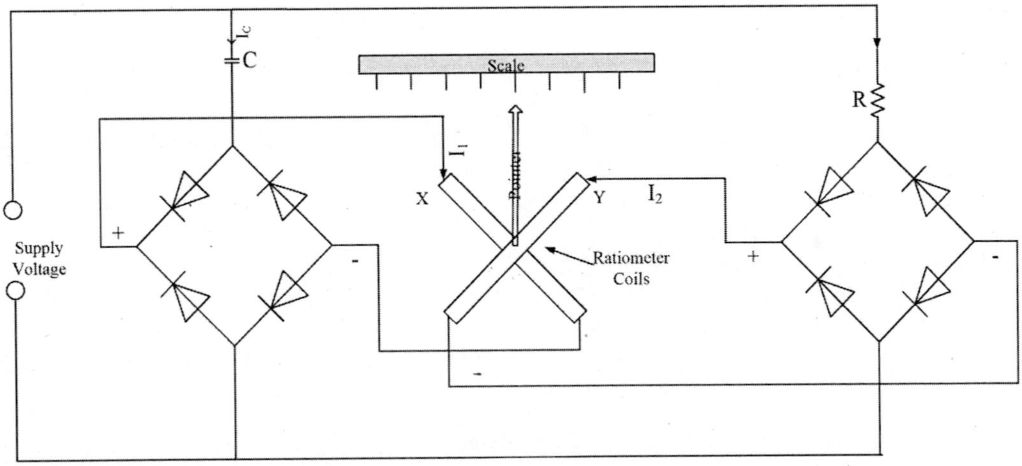

Fig. 5.13: Construction of an Electrodynamic Frequency Meter

Working:

When the frequency is connected to the supply, the rectified currents I_1 and I_2 pass through the moving coils C_1 and C_2. When these coils come to rest position, their torques become equal but in opposite directions. These meters depend on the value of resistance and a capacitor. These frequency meters have ranges of 40 to 60 Hz, 1200 to 2000 Hz, and 8000 to 12000 Hz.

5.6.3. Vibrating-reed Frequency Meter

A vibrating-reed frequency meter is a measuring instrument that is used to measure the frequency of various electric circuits. It consists of 7 vibrating reeds and each vibrating reed has a specific value. These reeds vibrate when this frequency meter is connected to the supply for the measurement of frequency. A reed that vibrates more as compared to the other reeds, the more vibrating reed is considered as frequency reading of a supply or electric circuit.

Construction:

This frequency meter consists of thin flat steel reeds which are arranged alongside and these reeds are near the electromagnet. The electromagnet consists of a laminated armature and a winding connected with a resistance in series on it. This winding is connected to the AC supply whose frequency is to be measured. The metallic reeds are 4 mm wide and 0.5 mm thick. These metallic reeds are arranged in a row and consist of a flexible base and these bases carry the armature of the electromagnet. These reeds are colored with white color and are arranged in ascending order of frequency on a scale.

Working:

To measure the frequency of the circuit, it is compulsory to connect the frequency meter to a supply. The electromagnet is connected to the supply for which frequency is to be measured. The magnetism of the electromagnet alternates with the same frequency and the electro-magnet exerts the attracting force on each reed once every half cycle. All reeds start vibrating but the reed whose frequency is double vibrates with maximum amplitude due to mechanical-resonance. The vibration of the other reeds is so small and these are unobservant. These vibrating reed frequency meters consist of small readings from 47 Hz to 53 Hz and 57 Hz to 63 Hz.

5.7. Sensors and Transducers

During measurement, the physical quantity under test makes its first contact with a sensor. A sensor is a device that produces an output signal to sense a physical phenomenon. In the broadest definition, a sensor is a device, module, machine, or subsystem that detects events or changes in its environment and sends the information to other electronics, frequently a computer processor. Sensors are always used with other electronics.

A transducer is a device that converts energy from one form to another. Usually, a transducer converts a signal in one form of energy to a signal in another. Transducers are often employed at the boundaries of automation, measurement, and control systems, where electrical signals are converted to and from other physical quantities (energy, force, torque, light, motion, position, etc.). The process of converting one form of energy to another is known as transduction. Table 5.1 provides the difference between a sensor and a transducer.

Table 5.1: Key difference between a sensor and a transducer

Sensor	Transducer
A sensor is a device that detects a change in a physical environment.	A transducer is a device that converts one form of energy into another.
A sensor is not necessarily a transducer.	Every transducer includes a sensor as a component.
A sensor itself is a component.	Transducer is made of a sensor and a signal conditioning circuit.
A sensor converts physical quantities or energy into a non-electrical signal.	A transducer converts physical quantity or energy into an electrical signal.
A sensor requires an additional circuit to process its output signal into a readable form.	A transducer does not require any processing circuit. Its output is directly interfaced with a device or display.
A sensor's output is analog.	A transducer can generate analog as well as digital output.
A sensor's output cannot be directly applied to any other system.	A transducer's output can be directly connected to another system.
A sensor does not require external power to operate.	A passive transducer requires an external power source to operate.
A sensor cannot be bidirectional i.e. it only converts physical quantities into readable form.	A transducer is bidirectional. It can also convert an electrical signal into physical quantities called an inverse transducer.
A sensor is a simple device.	A transducer has a complicated electrical circuit used for energy conversion.
Examples: Thermometer, pressure sensor, ultrasonic sensor, light sensor, etc.	Examples: Thermistor, potentiometer, piezoelectric transducer, Hall Effect transducer, etc.

5.7.1. Electrical Transducers

Electrical transducers are defined as transducers that convert one form of energy to electrical energy for measurement purposes. The quantities which cannot be measured directly, such as pressure, displacement, temperature, humidity, fluid flow, etc., are required to be sensed and changed into electrical signals first for easy measurement. The advantages of an electrical transducer are as-

1. An electrical signal obtained from an electrical transducer can be easily processed (mainly amplified) and brought to a level suitable for the output device which may be an indicator or recorder.
2. The electrical systems can be controlled with a very small level of power
3. The electrical output can be easily used, transmitted, and processed for measurement.
4. With the advent of IC technology, electronic systems have become extremely small in size, requiring small space for their operation.
5. No moving mechanical parts are involved in the electrical systems. Therefore there is no question of mechanical wear and tear and no possibility of mechanical failure.

5.7.2. Characteristics of a Transducer

- ***Dynamic range:*** This is the ratio between the largest amplitude signal and the smallest amplitude signal the transducer can effectively translate. Transducers with larger dynamic ranges are more "sensitive" and precise.
- ***Repeatability:*** This is the ability of the transducer to produce an identical output when stimulated by the same input.
- ***Noise:*** All transducers add some random noise to their output. In electrical transducers, this may be electrical noise due to the thermal motion of charges in circuits. Noise corrupts small signals more than large ones.
- ***Hysteresis:*** This is a property in which the output of the transducer depends not only on its current input but its past input.

5.7.3. Factors influencing the choice of transducers:

Some of the factors to be taken into consideration in the selection of a transducer for a particular application are,

1. ***Operating Principle:*** The first important factor for selecting a transducer is the operating principle. Various transducers use different operating principles like resistive, capacitive, piezoelectric, inductive, optoelectronic principle, etc.
2. ***Operating Range:*** The operating range is also one of the important factors to be considered because every instrument has its operating rating for satisfactory operation. The input applied must lie within this operating range so that the transducer can function with good resolution without any breakdown. The operating range of a transducer can be determined by its capabilities and error in the measurement.
3. ***Accuracy:*** The accuracy of a transducer is the degree of closeness of the output obtained to the true or ideal value. A transducer with high sensitivity can produce errors easily with other stimuli. The errors can be reduced using in-place system calibration and monitoring so that corrections are made accordingly to have a high degree of accuracy and repeatability.

4. ***Sensitivity:*** It is also a desirable selection factor of a transducer. Every transducer should be sufficiently sensitive to the input applied to produce an output. The sensitivity of a transducer is the output obtained per unit change in the input quantity. The sensitivity should not also be either very high or low which leads to errors.
5. ***Stability and Reliability***: A transducer should have high stability and reliability. It should be stable to external disturbances during its operation without deviating from the output. The stability of the transducer also describes storage life.
6. ***Usage and Ruggedness:*** A transducer should be mechanically rugged depending upon the application where it is used.
7. ***Transient Response and Frequency Response:*** The transducer should have required time domain specifications such as settling time, rise time, peak overshoot and small dynamic error, etc.
8. ***Loading Effects:*** The loading effect should be as minimum as possible so that errors in the measurement will be low. A transducer should have high input impedance and low output impedance for minimum loading effect.
9. ***Static Characteristics:*** The selected transducer should have low hysteresis, high linearity, and high resolution.
10. ***Environmental Compatibility:*** The factors of selection also include the consideration of environmental conditions. An incorrect choice of the transducer at a location to be operated where it can be subjected to temperature variation, vibration, and electromagnetic interference can affect the output.

5.7.4. Classification of Transducers:

The transducers can be classified based on:

(a) Transduction form used
(b) Primary and secondary transducers
(c) Passive and active transducers
(d) Analog and digital transducers
(e) Transducers and inverse transducers

(a) ***Classification based on the Principle of Transduction***
The transducer is classified by the transduction medium. The transduction medium may be resistive, inductive, or capacitive depending on the conversion process that is how the input transducer converts the input signal into resistance, inductance, and capacitance respectively.

(b) ***Primary and Secondary Transducer***
Primary Transducer – The transducer consists the mechanical as well as electrical devices. The mechanical devices of the transducer change the physical input quantities into a mechanical signal. This mechanical device is known as the primary transducer.

Secondary Transducer – The secondary transducer converts the mechanical signal into an electrical signal. The magnitude of the output signal depends on the input mechanical signal.

(c) ***Passive and Active Transducer***

Passive Transducer – The transducer which requires power from an external supply source is known as the passive transducer. Capacitive, resistive, and inductive transducers are an example of passive transducers.

Active Transducer – The transducer which does not require an external power source is known as the active transducer. Such type of transducer develops voltage or current on its own, hence known as a self-generating transducer. The output signal is obtained from the physical input quantity. The physical quantity like velocity, temperature, force, and intensity of light is induced with the help of the transducer. The piezoelectric crystal, photo-voltaic cell, tacho generator, thermocouples, and photovoltaic cell are examples of active transducers.

(d) ***Analog and Digital Transducer***

The transducer can also be classified by its output signals. The output signal of the transducer may be continuous or discrete.

Analog Transducer – The Analog transducer changes the input quantity into a continuous function. The strain gauges, L.V.D.T, thermocouple, and thermistor are examples of analog transducers.

Digital Transducer – These transducers convert an input quantity into a digital signal or in the form of a pulse. The digital signals work on high or low power.

(e) ***Transducer and Inverse Transducer***

Transducer – The device which converts the non-electrical quantity into an electric quantity is known as the transducer.

Inverse Transducer – The transducer which converts the electric quantity into a physical quantity, such type of transducer is known as the inverse transducer. The transducer has high electrical input and low non-electrical output.

5.8. Strain Gauges

If a metal conductor is stretched or compressed, its resistance changes due to the change in its length and diameter. There is also a change in the value of resistivity of the conductor when it is strained. This property is called the piezoresistive effect. Therefore, resistance strain gauges are also known as piezoresistive gauges. The strain gauges are used for the measurement of strain and associated stress in experimental stress analysis. Secondly, many other detectors and transducers, such as load cells, torque meters, diaphragm-type pressure gauges, temperature sensors, accelerometers, and flow meters, employ strain gauges as secondary transducers.

Theory of Strain Gauges

The change in the value of resistance by straining the gauge may be partly explained by the normal dimensional behavior of elastic material. If a strip of elastic material is positively strained, its longitudinal dimension will increase with a reduction in the lateral dimension. Hence, when a gauge is subjected to a positive strain, its length increases while its area of cross-section decreases. Since the resistance of a conductor is proportional to its length and inversely proportional to its area of cross-section, the resistance of the gauge increases with positive strain. The change in the value resistance of the strained conductor is more than what can be accounted for an increase in resistance due to dimensional changes. The extra change in the value of resistance is attributed to a change in the value of resistivity of a conductor when strained. This property is known as the piezoresistive effect.

Let us consider a strain gauge made of circular wire with the dimensions as-

Length=L, area= A, diameter=D before being strained. The material of the wire has a resistivity ρ.

∴ Resistance of unstrained gauge $R=\rho L/A$.

Let a tensile stress s be applied to the wire. This produces a positive strain causing the length to increase and the area to decrease. Thus when the wire is strained there are changes in its dimensions. Let ΔL=change in length, ΔA =change in the area, ΔD = change in diameter, and ΔR= change in resistance.

To find how ΔR depends upon the material physical quantities, the expression for R is differentiated to stress s.

$$\frac{dR}{ds}=\frac{\rho}{A}\frac{\partial L}{\partial s}-\frac{\rho L}{A^2}\frac{\partial A}{\partial s}+\frac{L}{A}\frac{\partial \rho}{\partial s} \quad (1)$$

Dividing Eq. 1 by resistance $R=\rho L/A$

$$\frac{dR}{ds}\frac{1}{R}=\frac{1}{L}\frac{\partial L}{\partial s}-\frac{1}{A}\frac{\partial A}{\partial s}+\frac{1}{\rho}\frac{\partial \rho}{\partial s} \quad (2)$$

It is evidenced from Eq. 2 that the per unit change in the resistance is due to-

a. per unit change in the length
b. per unit change in the area
c. per unit change in the resistivity

Now the Poisson's ratio is given as-

$$v=\frac{lateral\ strain}{longitudinal\ strain}=\frac{\partial D/_D}{\partial L/_L} \quad (3)$$

Or $\partial D/_D=-v\times\partial L/_L$ (4)

$$\frac{dR}{ds}\frac{1}{R}=\frac{1}{L}\frac{\partial L}{\partial s}-v\frac{2}{L}\frac{\partial L}{\partial s}+\frac{1}{\rho}\frac{\partial \rho}{\partial s} \quad (5)$$

For a small variation, the above equation can be written as-

$$\frac{\Delta R}{R}=\frac{\Delta L}{L}+2v\frac{\Delta L}{L}+\frac{\Delta \rho}{\rho} \quad (6)$$

The gauge factor is defined as the ratio of per unit change in resistance to per unit change in length.

Gauge factor $G_f=\frac{\Delta R/_R}{\Delta L/_L}$ (7)

Or $\frac{\Delta R}{R}=G_f\frac{\Delta L}{L}=G_f\times\varepsilon$ (8)

Where $\varepsilon=strain=\frac{\Delta L}{L}$

The gauge factor can be written as-

$$G_f = 1 + 2v + \frac{\Delta\rho/\rho}{\varepsilon} \quad (9)$$

$G_f =$ resistance change due to change in length + resistance change due to the change in the area + resistance change due to piezo electric effect

The strain is generally expressed in terms of micro strain. 1 micro strain =1μm/m. If the change in the value of resistivity of a material when strained is neglected, the gauge factor is-

$$G_f = 1 + 2v \quad (10)$$

Eq. 10 is valid only when the piezoresistive effect is almost negligible. Table 5.2 gives the gauge factors for the various materials.

Table 5. 2 Gauge Factors

Material	Gauge factors	Material	Gauge factors
Nickel	- 12.1	Platinum	+ 4.8
Manganin	+ 0.47	Carbon	+ 20
Nichrome	+ 2.0	Doped Crystals	100-5000
Constantan	+ 2.1		
Soft iron	+ 4.2		

5.9. Thermistors

A thermistor is a semiconductor type of resistor whose resistance is strongly dependent on temperature, more so than in standard resistors. Depending on the materials used, thermistors are classified into two types:

- With NTC thermistors, resistance decreases as temperature rises; usually due to an increase in conduction electrons bumped up by thermal agitation from the valence band. An NTC is commonly used as a temperature sensor, or in series with a circuit as an inrush current limiter.
- With PTC thermistors, resistance increases as temperature rises; usually due to increased thermal lattice agitations particularly those of impurities and imperfections. PTC thermistors are commonly installed in series with a circuit and used to protect against overcurrent conditions, as resettable fuses.

Construction: Thermistors are composed of a sintered mixture of metallic oxides such as manganese, nickel, Cobalt, copper, iron, and uranium. They are available in a variety of sizes and shapes. The sensing element is covered with an insulating material before enclosing it with the metal tube. Two leads are connected to the temperature-sensing element and are brought out of the metal tube. The other end of the two leads is connected to one of the arms of the bridge circuit (generally Wheatstone bridge is used) which measures the resistance of the temperature-sensing element. The thermistors may be in the form of beads, rods, or discs. Commercial forms are shown in Fig. 5.14.

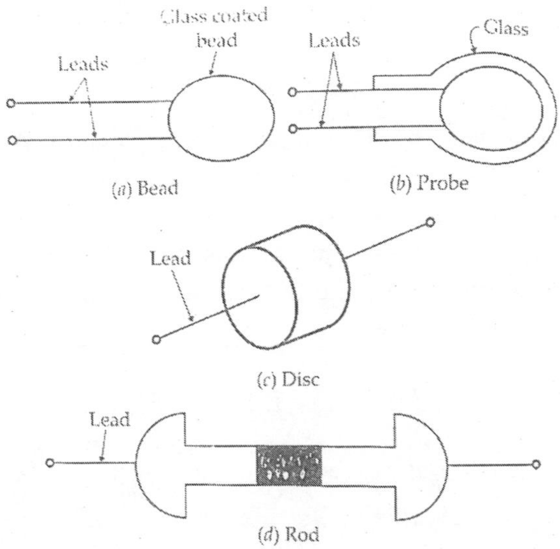

Fig.5.14: Types of Thermistors

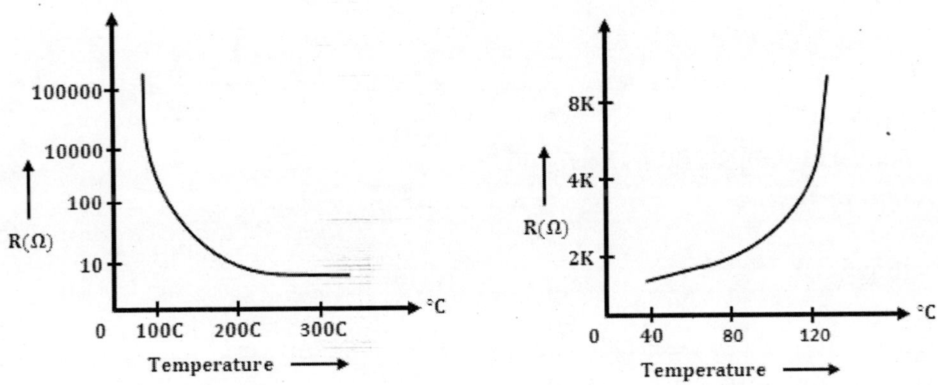

Resistance Temperature Characteristics of NTC Thermistor

Resistance Temperature Characteristics of PTC Thermistor

Fig. 5.15: Resistance- Temperature Characteristics

The mathematical expression for the relationship between the resistance of a thermistor and the absolute temperature of a thermistor is

$$R_{T1} = R_{T2}\, e^{\left[\beta\left(\frac{1}{T_1}-\frac{1}{T_2}\right)\right]}$$

Where-

R_{T1}= resistance of the thermistor at absolute temperature T_1; K

R_{T2}= resistance of the thermistor at absolute temperature T_2; K β= a constant depending upon the material of the thermistor, typically 3500 to 4500 K.

The resistance temperature characteristics of Fig. 5.15 show that a thermistor has a very high negative temperature co-efficient of resistance, making it an ideal temperature transducer. The characteristics of thermistors are no doubt nonlinear but a linear approximation of the resistance temperature curve can be obtained over a small range of temperatures. Thus for a limited range of temperatures, the resistance of a thermistor varies as given by the equation-

$$R_\theta = R_{\theta 0}[1 + \alpha_{\theta 0}\Delta\theta]$$

A thermistor exhibits a negative resistance temperature co-efficient, which is typically 0.05Ω/Ω -°C. In place of linear approximation, an approximate logarithmic relationship may be used for the resistance-temperature relationship for a thermistor. The relationship is given by the equation-

$$R_T = aR_0 e^{\frac{b}{T}}$$

where

R_θ = resistance at ice point (Ω),
Ro = resistance at temperature T(K), and
a, b are constants.

Applications:

1. It is used for the measurement of high-frequency power.
2. The thermistor measures thermal conductivity.
3. The thermistor measures the pressure of the liquid.
4. It measures the composition of gases.
5. The thermistor measures the vacuum and provides the time delays.
6. It is used in temperature compensation applications
7. It is used for temperature measurement and control applications

5.10. Linear variable differential transformer (LVDT)

The linear variable differential transformer (LVDT) (also called linear variable displacement transformer is a type of electrical transformer used for measuring linear displacement (position). It is the most widely used inductive transducer that converts linear motion into an electrical signal. The output across the secondary of this transformer is the differential thus it is called so.

Construction:

Main Features of Construction

- The transformer consists of a primary winding P and two secondary windings S1 and S2 wound on a cylindrical former (which is hollow and contains the core).
- Both the secondary windings have an equal number of turns, and we place them on either side of the primary winding
- The primary winding is connected to an AC source which produces a flux in the air gap and voltages are induced in secondary windings.
- A movable soft iron core is placed inside the former and the displacement to be measured is connected to the iron core.
- The iron core is generally of high permeability which helps in reducing harmonics and high sensitivity of LVDT.
- The LVDT is placed inside a stainless steel housing because it will provide electrostatic and electromagnetic shielding.
- Both the secondary windings are connected in such a way that resulted output is the difference between the voltages of the two windings.

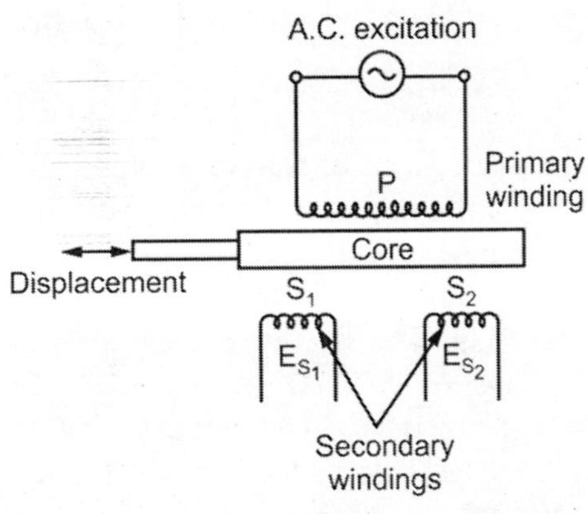

Fig. 5.16: The construction of LVDT

The Working:

As the primary is connected to an AC source so alternating current and voltages are produced in the secondary of the LVDT. The output in secondary S_1 is e_1 and in secondary S_2 is e_2. So the differential output is

$$e_{out} = e_1 - e_2$$

This equation explains the **principle of operation of LVDT.**

Three cases arise according to the locations of the core which explains the working of LVDT are discussed below as,

- CASE I When the core is at null position (for no displacement): When the core is at a null position then the flux linking with both the secondary windings is equal so the induced emf is equal in both the windings. So for no displacement, the value of output e_{out} is zero as e_1 and e_2 both are equal. So it shows that no displacement took place.
- CASE II When the core is moved to the upward of null position (For displacement to the upward of reference point):
  In this case, the flux linking with secondary winding S_1 is more as compared to flux linking with S_2. Due to this e_1 will be more than that e_2. Due to this output voltage, e_{out} is positive.
- CASE III When the core is moved to the downward Null position (for displacement to the downward of the reference point):
  In this case, the magnitude of e_2 will be more than that of e_1. Due to this output e_{out} will be negative and shows the output downward of the reference point.

The magnitude and sign of voltage induced in LVDT are related as-

- The amount of change in voltage either negative or positive is proportional to the amount of movement of the core and indicates the amount of linear motion.
- By noting the output voltage increasing or decreasing the direction of motion can be determined
- The output voltage of an LVDT is a linear function of core displacement.

Advantages of LVDT

- High Range – The LVDTs have a very high range for the measurement of displacement. They can be used for the measurement of displacements ranging from 1.25 mm to 250 mm
- No Frictional Losses – As the core moves inside a hollow former so there is no loss of displacement input as frictional loss so which makes LVDT a very accurate device.
- High Input and High Sensitivity – The output of LVDT is so high that it doesn't need any amplification. The transducer possesses a high sensitivity which is typically about 40V/mm.
- Low Hysteresis – LVDTs show a low hysteresis and hence repeatability is excellent under all conditions
- Low Power Consumption – The power is about 1W which is very less as compared to other transducers.
- Direct Conversion to Electrical Signals – They convert the linear displacement to an electrical voltage which is easy to process

Disadvantages of LVDT

- LVDT is sensitive to stray magnetic fields so it always requires a setup to protect them from stray magnetic fields.
- LVDT gets affected by vibrations and temperature.

Applications of LVDT

1. LVDT can be used in applications where displacements to be measured are ranging from a fraction of millimeters to a few centimeters. The **LVDT** acting as a primary transducer converts the displacement to an electrical signal directly.
2. The LVDT can also act as a secondary transducer. E.g. the Bourbon tube acts as a primary transducer and it converts pressure into linear displacement and then LVDT coverts this displacement into an electrical signal which after calibration gives the readings of the pressure of the fluid.

5.11. Hall Effect Transducer

The Hall Effect element is a type of transducer used for measuring the magnetic field by converting it into an emf. The direct measurement of the magnetic field is not possible. Thus, the Hall Effect Transducer is used for indirect measurement of the magnetic field. The transducer converts the magnetic field into an electric quantity which is easily measured by the analog and digital meters and is called a Hall Effect transducer.

The working principle of the Hall Effect Transducer is based on Hall Effect. It was discovered by Edwin Hall in 1879. Hall Effect is the process of development of potential difference across the two faces of a current-carrying strip when the strip is kept in a magnetic field.
The magnitude of voltage depends upon the current, strength of the magnetic field, and the property of conducting material. The Hall Effect is found in conducting material and semiconductors in varying amounts depending upon the density and mobility of the current carrier. The Hall Effect elements are based on a thin film of semiconducting material like indium arsenide.

Working:

Fig. 5.17 shows the concept of the Hall Effect transducer and its working. The current supply is through the lead 1 and 2 and the output is obtained from strips 3 and 4. Lead 3 and 4 are at the same potential when no field is applied across the strip. When the magnetic field is applied to the strip, the output voltage develops across the output leads 3 and 4. This developed voltage is directly proportional to the strength of the material.

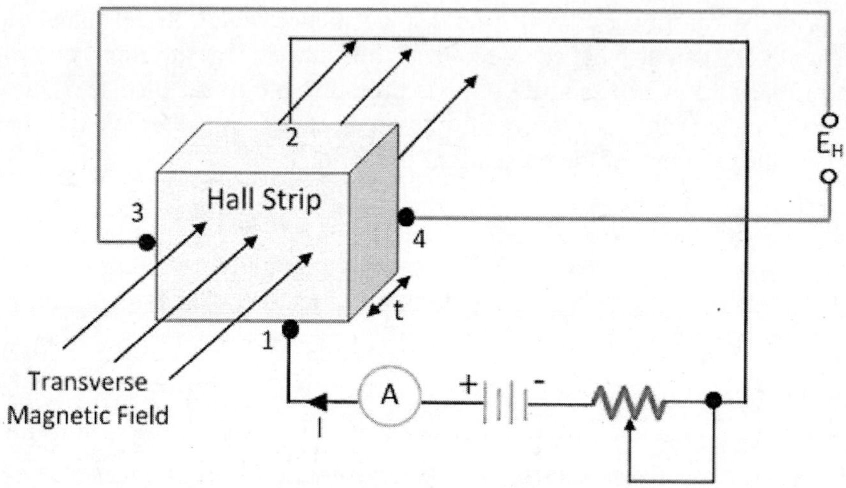

Fig. 5.17: Hall Effect Element

The voltage is given by

$$V_H = K_H \frac{IB}{t}$$

Where- K_H = Hall effect coefficient

I = Current

B = magnetic flux density

t = thickness of the strip (hall element)

The emf produced is called hall voltage and depends upon the current, flux density, and a property of the conductor known as the Hall Effect coefficient (K_H).

Advantages:

- It has a long life because of no wear.
- They are highly reliable.
- It has a very high-speed operation.
- They can be operated in a wide temperature range.

Disadvantages:

- It cannot measure a distance of more than 10cm.
- A stray magnetic field can cause an inaccurate reading.

Applications:

- The Hall Effect element is used for converting the magnetic flux into an electric transducer.
- The Hall Effect element measures the displacement of the structural element.
- The Hall Effect transducer is also used for measuring the current without any physical connection between the conductor circuit and the meter.
- The Hall Effect transducer is used for measuring the power of the conductor. The current is applied across the conductor, which develops the magnetic field.
- Open/close of laptop flip screen. Hence conserve power while switching the laptop to sleep.
- Used in encoded switches and rotary encoders.

5.12: Capacitive Transducer

The capacitive transducer is used for measuring displacement, pressure, and other physical quantities. It is a passive transducer which means it requires external power for operation. The capacitive transducer works on the principle of variable capacitances. The capacitance of the capacitive transducer changes because of many reasons like overlapping of plates, change in distance between the plates, and dielectric constant.

The capacitive transducer contains two parallel metal plates. These plates are separated by the dielectric medium which is either air, material, gas, or liquid. In the normal capacitor, the distance between the plates is fixed, but in the capacitive transducer, the distance between them is varied.

The capacitive transducer uses the electrical quantity of capacitance for converting the mechanical movement into an electrical signal. The input quantity causes the change of the capacitance which is directly measured by the capacitive transducer. The capacitors measure both static and dynamic changes. The displacement is also measured directly by connecting the measurable devices to the movable plate of the capacitor. It works with both the contacting and non-contacting modes.

Working:

The equations below express the capacitance between the plates of a capacitor

$$C = \frac{\varepsilon A}{d} = \frac{\varepsilon_r \varepsilon_o A}{d}$$

Where A=overlapping area of plates in m^2

d=the distance between two plates in meter
ε= permittivity of the medium in F/m
ε_r= relative permittivity
ε_0 = the permittivity of free space
The schematic diagram of a parallel plate capacitive transducer is shown in Fig. 5.18.

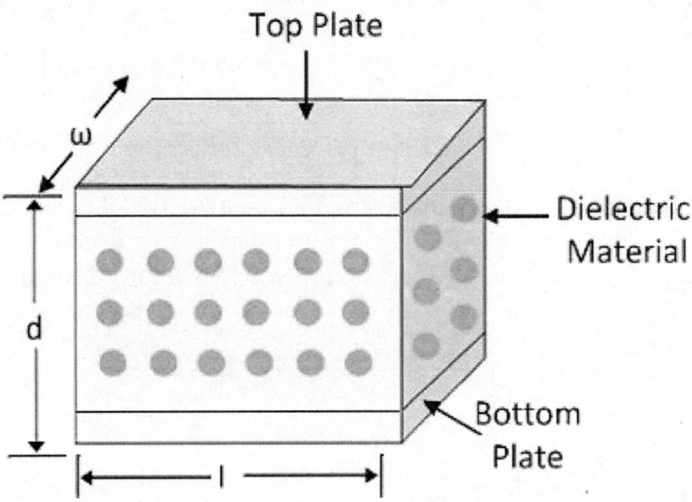

Fig. 5.18: Parallel plate capacitive transducer

The change in capacitance occurs because of physical variables like displacement, force, pressure, etc. The capacitance of the transducer also changes by the variation in their dielectric constant which is usually because of the measurement of liquid or gas level. The capacitance of the transducer is measured with the bridge circuit. The output impedance of the transducer is given as

$$X_C = \frac{1}{2\pi f_c}$$

Where, C – capacitance; f – frequency of excitation in Hz.

The capacitance between two plates can be varied by any of the following methods.

- By changing the distance between two plates (**d**)
- By changing the permittivity of the dielectric medium (ϵ)
- By changing the area of overlapping of plates (**A**)

By changing the distance between two plates: The capacitance can be varied by changing the distance between two plates. From the equation for C, we can observe that C and d are inversely proportional to each other. That is, the capacitance value will decrease with increasing distance and vice-versa. This principle can be used in a transducer by making the left plate fixed and the right plate movable by the displacement that is to be measured as shown in Fig. 5.19.

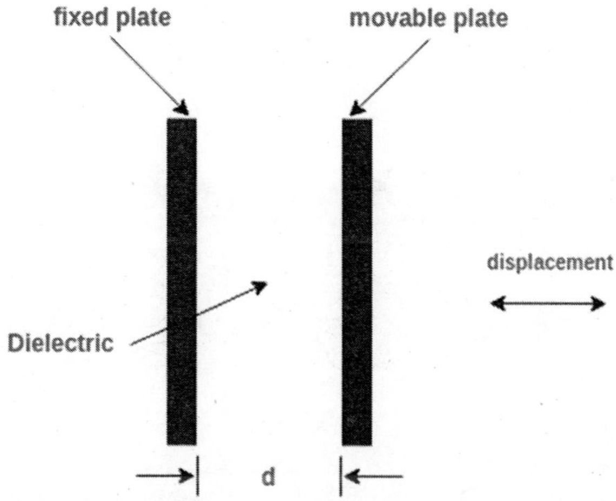

Fig. 5.19: Variation in the distance between two plates

By changing the permittivity of the dielectric medium

Another method to change the capacitance value is by changing the permittivity of the dielectric material (ϵ). The permittivity and capacitance values are directly proportional to each other.

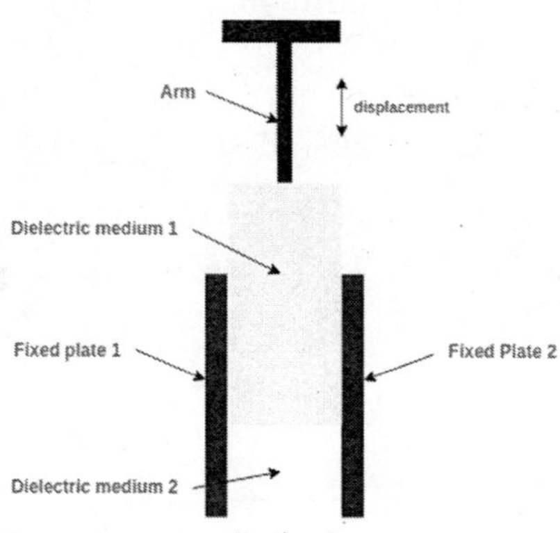

In this arrangement, a dielectric material is filled into the space between the two fixed plates. It can be moved using the arm. This causes a variation in the dielectric constant in the region. The change in dielectric constant will vary the capacitance of the transducer.

By changing the area of overlapping plates

The capacitance can also be changed by varying the area of overlapping plates. As shown in Fig. 5.20, one plate is kept fixed, and the other is movable. When the plate is moved, the area of overlapping of plates changes, and the capacitance also changes. The capacitance value and area are directly proportional to each other. These types of transducers are used to measure relatively large displacements.

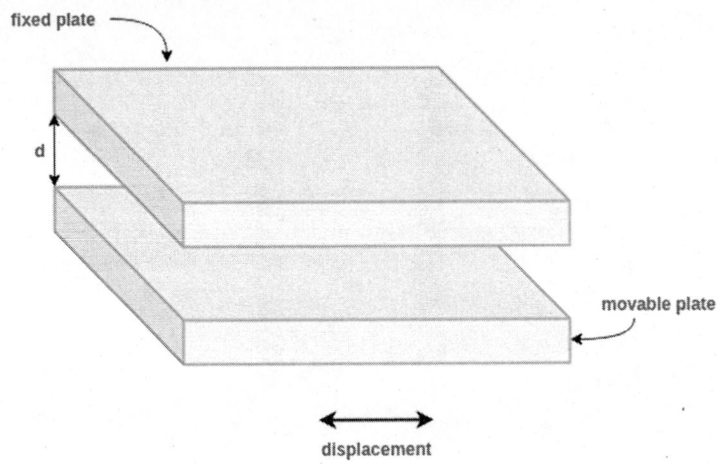

Fig.5.20: Capacitance between two plates

Advantages:

1. It requires an external force for operation and is hence very useful for small systems.
2. The capacitive transducer is very sensitive.
3. It gives a good frequency response which is used for the dynamic study.
4. The transducer has high input impedance hence they have a small loading effect.
5. It requires small output power for operation.

Disadvantages:

1. The metallic parts of the transducers require insulation.
2. The frame of the capacitor requires earthling for reducing the effect of the stray magnetic field.
3. Sometimes the transducer shows nonlinear behaviors because of the edge effect which is controlled by using the guard ring.
4. The cable connecting across the transducer causes an error.

Applications:

1. The capacitive transducer uses for the measurement of both the linear and angular displacement. It is extremely sensitive and used for the measurement of very small distances.
2. It is used for the measurement of force and pressure. The force or pressure, which is to be measured is first converted into a displacement, and then the displacement changes the capacitances of the transducer.
3. It is used as a pressure transducer in some cases, where the dielectric constant of the transducer changes with the pressure.
4. The humidity in gases is measured through the capacitive transducer.
5. The transducer uses the mechanical modifier for measuring the volume, density, weight, etc.

5.13. Piezoelectric Transducer

A piezoelectric transducer is an instrument that applies the piezoelectric effect to measure variations in strain, temperature, acceleration, pressure, or force by transforming this energy into an electrical load. The electric voltage generated by a piezoelectric transducer can be simply evaluated by the devices of voltage measuring. This voltage is an operation of the pressure or force influenced by it.

A piezoelectric transducer uses a piezoelectric material as a transduction element. A piezoelectric material is one in which an electrical potential difference appears across a certain surface of a crystal if the dimension of the crystal is changed by the application of force. This potential difference appears due to displacement of charge. This effect is known is Piezoelectric Effect. Rochelle Salt, Ammonium Dihydrogen Phosphate, Lithium Suphate, dipotassium tartarate, quartz, and ceramic are some common examples of piezoelectric material.

There are two types of piezoelectric materials: Natural and Synthetic. A natural piezoelectric material occurs in nature and can be used as such. However, synthetic piezoelectric materials are those materials in which piezoelectric properties are not found in their original state but these properties are produced using special techniques such as polarizing treatment. Quartz and Ceramic are examples of natural piezoelectric materials whereas materials like lithium sulphate, and ethylene diamine tartarate belong to the synthetic group.

Working Principle:

The working principle of a Piezoelectric Transducer is based on the fact that when a mechanical force is applied to a piezoelectric crystal, a voltage is produced across its faces. Thus, mechanical phenomena are converted into electrical signals. No external supply is required for this transducer to work and hence it is an active transducer.

Piezoelectric Transducer responds to the mechanical force/deformation and generates voltage. There may be various modes of deformation to which these transducers can respond. The modes can be thickness expansion, transverse expansion, thickness-shear, and face shear.

In a piezoelectric transducer, a piezoelectric crystal is sandwiched between the two electrodes. When a mechanical deformation takes place, it generates charge and hence it acts as a capacitor. A voltage is developed across the electrodes of the transducer which can be measured and calibrated with the deforming force to directly measure the mechanical deforming force. Fig.5.21 shows a simple piezoelectric transducer.

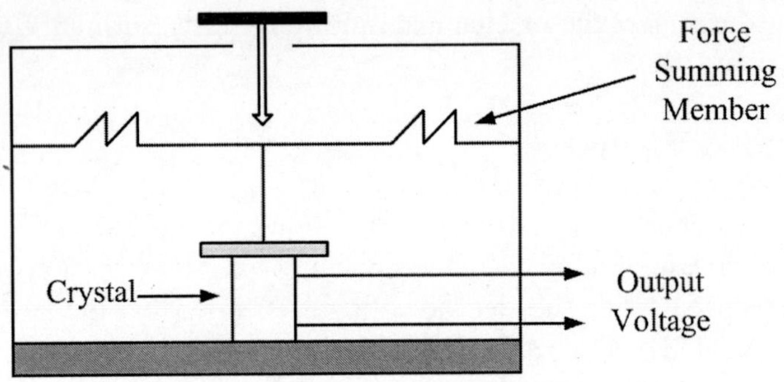

Fig. 5.21: Piezoelectric Transducer

The piezoelectric effect is direction sensitive i.e. the polarity of charge will not be the same for a tensile and compressive force. The polarity of voltage induced due to a tensile force will be opposite to the polarity of voltage produced due to a compressive force.

The magnitude and polarity of the induced charge on the electrodes are directly proportional to the applied force and its direction.

Let the applied force be F, then the charge induced will be given as

Q = kF ....(1)

where k is the constant of proportionality. This constant is the charge sensitivity of the piezoelectric material. It is constant for a given material and defined as the charge generated per unit applied force. Its unit is (Columb / Newton)

Assuming the surface area of the electrode, and separation between the electrodes be **A** and **d** respectively, the charge generated on each of the electrodes of the piezoelectric transducer is given below.

$Q = CV$

where C is capacitance formed by the electrodes and the piezoelectric material

$C = \varepsilon A / d$

Therefore,

$Q = \varepsilon AV / d$ ....(2)

From (1) and (2),

$kF = \varepsilon AV / d$

$F = (\varepsilon AV) / (dk)$

Where, ε, A, d, and k are constant for a given piezoelectric transducer. This essentially means that the magnitude of applied force is directly proportional to the output voltage across the electrodes.

Thus, by measuring the value of voltage across the electrodes of a piezoelectric transducer, the value of mechanical force can be calculated. Hence, mechanical force is converted into an electrical signal.

The applications of a piezoelectric material and transducer are listed below:

1. Quartz is commonly used for stabilizing electronic oscillators due to its high stability.
2. Piezoelectric Transducer is mainly used in dynamic measurements. The voltage developed by the application of strain is not held under static conditions. Therefore, these transducers are used in the measurement of quantities such as Surface Roughness, acceleration (called accelerometer), and vibrations.
3. The ultrasonic generator also uses Barium titanate which is a piezoelectric material. Such materials are used in industrial cleansing apparatus and also in underwater detection systems known as sonar.

5.14. Seismic Transducer

The seismic transducer is used for measuring the vibration of the ground. The spring-mass damper element and the displacement transducer are the two main components of the seismic transducer.
The mass that is connected to the damper element and spring without any other support is known as the spring mass damper element. And the displacement transducer converts the displacement into the electrical quantity. The seismic transducer is used for measuring the earth's vibration, volcanic eruption, and other vibrations, etc.
The systematic diagram of the seismic transducer is shown in Fig. 5.22. The mass is connected with the help of the damper and spring to the housing. The housing frame is connected to the source whose vibrations need to be measured.

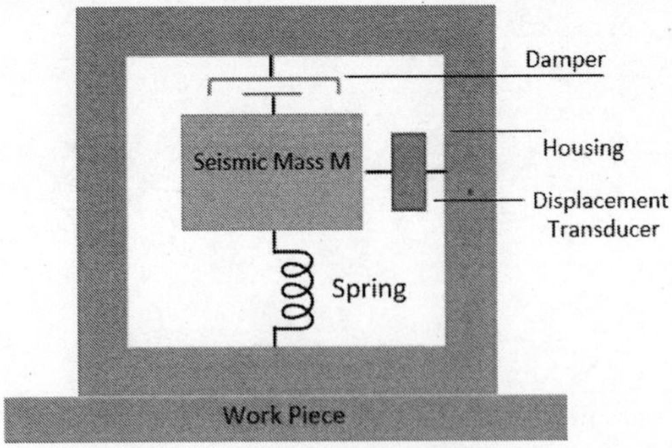

Fig. 5.22: Construction of a Seismic Transducer

The arrangement is kept in such a way that the position of the mass remains the same in the space. Such type of arrangement is kept for causing the relative motion between the housing frame and the mass. The term relative motion means one of the objects remains stationary, and the other is in motion concerning the first one. The displacement that occurs between the two is sensed and represented by the transducer.

The seismic transducer works in two different modes.

- Displacement Mode
- Acceleration Mode

The selection of the mode depends on the combinations of the mass, spring, and damper combinations. The large mass and soft spring are used for the displacement mode measurement while the combination of the small mass and stiff spring is used for the acceleration mode.

Types of Seismic Transducer

Vibrometer and the accelerometer are the two types of seismic transducers.

1. Vibrometer – The vibrometer or low-frequency meter is used for measuring the displacement of the body. It also measures the high frequency of the vibrating body. Their frequency range depends on the natural frequency and the damping system.

2. Accelerometer – The accelerometer measures the acceleration of the measuring body. The acceleration shows the total force acting on the object.

*****

Solved examples

Ex. 1. A resistance wire strain gauge uses a soft iron wire of a small diameter. The gauge factor is +4.2. Neglecting the piezoresistive effects, calculate the Poisson's ratio.

Solution:
The gauge factor is calculated by-

$$G_f = 1 + 2v + \frac{\Delta\rho/\rho}{\varepsilon}$$

Neglecting the piezoelectric effect, the gauge factor can be written as-

$$G_f = 1 + 2v$$

Hence, Poisson's ratio $v = \frac{G_f - 1}{2} = \frac{(4.2-1)}{2} = 1.6$

Ex. 2. A compressive force is applied to a structural member. The strain is 5 micro-strain. Two separate strain gauges are attached to the structural member, one is a nickel wire strain gauge having a gauge factor of – 12.1 and the other is a Nichrome wire strain gauge having a gauge factor of 2. Calculate the value of resistance of the gauges after they are strained. The resistance of strain gauges before being strained is 120Ω.

Solution:
According to our convention, the tensile strain is taken as positive while the compressive strain is taken as negative.
Hence, strain $\varepsilon = -5 \times 10^{-6}$ (1 micro strain= 1 μm/m)
Now $\Delta R/R = G_f\varepsilon$

Hence, the change in the resistance of the nickel wire strain gauge-

$$\Delta R = G_f\varepsilon \times R$$

$$= (-12.1) \times (-5 \times 10^{-6}) \times 120$$

$$= 7.26 \times 10^{-3}\Omega = 7.26m\Omega$$

Thus there is an increase of 7.26mΩ in the value of the resistance.

For Nicrome, the change in the value of the resistance is-

$$\Delta R = (2) \times (-5 \times 10^{-6}) \times 120$$

$$= 1.2 \times 10^{-3}\Omega = -1.2m\Omega$$

Thus with compressive strain, the value of resistance decreases to 1.2mΩ.

Ex. 3. A thermistor has a resistance temperature coefficient of - 5 % over a temperature range of 25 to 50 ^{0}C. If the resistance of the thermistor is 100Ω at 25°C, what is the resistance at 35°C?

Solution:

Resistance at a temperature of 35^{0}C is :

$$R_\theta = R_{\theta 0}[1 + \alpha_{\theta 0}\Delta\theta]$$

R_{35} = 100[1-0.05(35-25)]

= 50Ω

Ex. 4. An automatic temperature control arrangement for an electric oven is having a thermistor with a resistance of 2100Ω at 70°C. A potentiometer is used for obtaining balance conditions. The temperature of the oven changes and therefore the rheostat has to be set at 1900°C to obtain balanced conditions. Using the linear approximation for the resistance temperature curve obtain the change in the value of resistance. The resistance temperature coefficient can be assumed as -0.04 Ω/Ω -°C.

Solution:

Using linear approximation equation:

1900=2100[1-0.04(θ-70)]

θ =67°C

Ex. 5: The output voltage of an LVDT is 1.2 V at maximum displacement. At a load of 0.4MΩ, the deviation from linearity is maximum and it is ±0.006 V from a straight line through the origin. Find the linearity at the given load.

Solution:

$$\%\,Linearity = \pm\frac{0.006}{1.2}\times 100$$

= ± 0.5%

Ex. 6. The output of an LVDT is connected to a 5 V voltmeter through an amplifier whose amplification factor is 250. An output of 2 mV appears across the terminals of LVDT when the core moves through a distance of 0.5 mm. Calculate the sensitivity of the LVDT and that of the whole setup. The millivolt meter scale has 100 divisions. The scale can be read to 1/5 of a division. Calculate the resolution of the instrument in mm.

Solution:

$$\text{Sensitivity of LVDT} = \frac{\text{output voltage}}{\text{Displacement}} = \frac{2\times 10^{-3}}{0.5} = 4\times 10^{-3}\text{V/ mm} = 4\text{ mV/mm}$$

Sensitivity of instrument = amplification factor × sensitivity of LVDT

$$= 4\times 10^{-3}\times 250 = 1\text{ V/mm.} = 1000\text{ mV/mm.}$$

Scale division =5/100 V=50 mV.

Minimum voltage that can be read on the voltmeter $=\frac{1}{5}\times 50 = 1$ mV.

Resolution of instrument $=1\times \frac{1}{1000} = 1\times 10^{-3}$ mm.

Ex.7. A steel cantilever is 0.25 m long, 20 mm wide, and 4 mm thick.
(a) Calculate the value of deflection at the free end for the cantilever when a force of 25 N is applied at this end. The modulus of elasticity for steel is 200 GN/m^2.
An LVDT with a sensitivity of 0.5 V/mm is used. The voltage is read on a 10 V voltmeter having 100 divisions. Two-tenths of a division can be read with certainty.
(b) Calculate the minimum and maximum value of force that can be measured with this arrangement.

Solution:

(a) Moment of the area of cantilever $I = \frac{1}{12} bt^3$
$= \frac{1}{12} \times (0.02) \times (0.004)^3 = 0.107\times 10^{-9} m^4$

Deflection $X = \frac{F^3}{3EI} = \frac{25\times(0.25)^3}{3\times 200\times 10^9 \times 0.107\times 10^{-9}} = 6.08\times 10^{-8}\ m = 6.08mm$

(b) Deflection per unit force $\frac{x}{F} = \frac{6.08}{25} = 0.2432$ mm/N

The overall sensitivity of the measurement system $= 0.2432\frac{mm}{N} \times 0.5\frac{V}{mm} = 0.1216$ V/N
1 scale division $= \frac{10}{1000} = 0.1$ V. Since two-tenths of a scale division can be read
With certainty, resolution $= \frac{2}{10} \times 0.1 = 0.02$V.

The minimum force that can be measured $= \frac{0.02}{0.1216} = 0.1645$ N.
The maximum force that can be measured $= \frac{10}{0.1216} = 82.2$ N.

Exercise Questions

1. With a neat diagram, explain the working principle of the PMMC device.
2. Describe different types of ohmmeters in detail.
3. Explain the working of a galvanometer.
4. Explain the working principle of a potentiometer in detail.
5. Discuss the principle of the D'Arsonval movement
6. Discuss the classification of the measuring instruments.
7. List and explain different types of frequency meters.
8. What is a transducer? Discuss the classification of the transducers.
9. List and explain the factors influencing the choice of transducers.
10. Explain the working principle of strain gauge.
11. Discuss LVDT in detail.
12. What is the working principle of a capacitive transducer?
13. Explain the construction and working of thermistors.
14. What is Hall Effect? Explain the construction and working of a Hall Effect transducer.
15. Explain the piezoelectric transducers in detail.
16. What is a Seismic transducer? Explain its classification and working.

Multiple Choice Questions

1. Which one of the following moving coil instruments is used only for DC?
   (a) PMMC
   (b) Dynamometer
   (c) Both a and b
   (d) None of the above
2. Which one of the following is a type of indicating instrument?
   (a) X-Y plotters
   (b) Watt-hour meter
   (c) Wattmeters and ammeters
   (d) All of the above
3. The deflection torque in moving coil instrument is proportional to ________?
   (a) Current
   (b) Voltage
   (c) Square of the current
   (d) None of the above
4. Which of the following is the most sensitive detector for single frequency value?
   (a) oscillator
   (b) headphone
   (c) tuned detector
   (d) vibration galvanometer
5. The unknown capacitance value is obtained by _________
   (a) using a vibration galvanometer
   (b) using the capacitance of other ratio arms
   (c) comparison with standard
   (d) using a tuned detector
6. Which of the following device is used to measure power in A.C. circuits?
   (a) ammeter
   (b) wattmeter
   (c) voltmeter
   (d) ohmmeter
7. The function of a transducer is to convert______
   (a) Electrical signal into non-electrical quantity
   (b) Nonelectrical quantity into an electrical signal
   (c) Electrical signal into mechanical quantity
   (d) All of these
8. Potentiometer transducers are used for the measurement of
   (a) Pressure
   (b) Displacement
   (c) Humidity
   (d) Both (a) and (b)
9. The strain gauge is
   (a) Active device and converts mechanical displacement into a change of resistance
   (b) Passive device and converts electrical displacement into a change of resistance
   (c) Passive device and converts mechanical displacement into a change of resistance

(d) Active device and converts electrical displacement into a change of resistance

10. For the measurement of pressure, the instruments used can be
    (a) Mechanical
    (b) Electro-mechanical
    (c) Electronic
    (d) All of these

11. The method/methods suitable for the measurement of low resistance is/are
    (a) Ammeter-voltmeter method
    (b) Kelvin's double bridge method
    (c) Potentiometer method
    (d) All of these

12. The readings of the two-wattmeter used for the measurement of power input to a 3-phase induction motor are 850 W and 250 W respectively. The power factor of the motor is
    (a) 0.73
    (b) 0.76
    (c) 0.79
    (d) 0.85

13. The deflection θ is related to the electric current I in a galvanometer by the relation
    (a) $I \propto \theta$
    (b) $I \propto \tan \theta$
    (c) $I \propto \sin \theta$
    (d) $I \propto \cos \theta$

14. The deflection in moving coil galvanometer is
    (a) Inversely proportional to the area of the coil
    (b) Directly proportional to the torsional constant
    (c) inversely proportional to the current flowing
    (d) Directly proportional to the number of turns of the coil

15. **_______ Transducer is used for measurement of the magnetic field.**
    (a) Hall effect
    (b) Inductive
    (c) LVDT
    (d) hygrometer

16. **Which of the following is not a material used in the Hall Effect sensor?**
    (a) Caesium-Antimony
    (b) Gallium-Arsenide
    (c) Indium-Antimonide
    (d) Indium-Arsenide

17. **The force exerted by the magnetic field in Hall Effect transducers is ____________.**
    (a) Lorentz force
    (b) Hall Effect force
    (c) Magnetic force
    (d) Electric force

18. **Hall Effect transducer can measure ______ current.**

(a) AC
(b) DC
(c) AC or DC
(d) Eddy

19. **The difference between a normal capacitor and a capacitive transducer is------**

(a) Area
(b) Intensity
(c) The distance between the plates
(d) Both a & b

20. **How the plates in the capacitive transducer are connected?**

(a) Both are fixed
(b) One is fixed and the other is movable
(c) Both are movable
(d) As per requirement

21. **Sensitivity in capacitive transducers is ______?**

(a) Low
(b) High
(c) Constant
(d) Unaltered

22. **Which of the following is a practical application of thermistors:**

(a) Fire alarm
(b) Automotives
(c) Food handling and processing
(d) All of these

23. **Which of the following is correct for thermistors:**

(a) A thermistor is a resistance thermometer
(b) Resistance of NTC thermistor increases with increasing temperature
(c) Resistance of PTC thermistor decreases with increasing temperature
(d) Both 2 & 3

24. **The temperature range of thermistor when compared to RTD is:**

(a) Smaller
(b) Same as RTD
(c) Larger
(d) None of above

25. **Thermistors can be considered as which type of Resistor:**

(a) Laser resistor
(b) Photo resistor
(c) Thermal resistor
(d) None of these

26. **The ___________ of a strain gauge varies with applied strain.**

(a) resistance
(b) capacitance
(c) inductance
(d) flux

27. **The bonding element in a strain gauge must have __________**

(a) zero insulation resistance
(b) low insulation resistance
(c) high insulation resistance
(d) infinite insulation resistance

28. **------------- device can be used for measurement of frequency?**

(a) Voltmeter
(b) Ammeter
(c) Stroboscope
(d) None of the mentioned

29. **___________ helps in current measurement by placing it in ___________ with the circuit element.**

(a) Voltmeter, Parallel
(b) Ammeter, series
(c) Voltmeter, series
(d) Ammeter, parallel

30. **An ideal voltmeter functions as _________ circuit.**

(a) A short
(b) An open
(c) A power
(d) An infinite

31. **The accuracy of D'Arsonval movements used in common laboratory meters is about ------------of the full-scale reading**

(a) 10%
(b) 1%
(c) 0.1%
(d) 5%

MCQ Answer key:

1	**a**	9	**c**	17	**a**	25	**c**
2	**a**	10	**d**	18	**c**	26	**a**
3	**c**	11	**d**	19	**c**	27	**c**
4	**c**	12	**a**	20	**b**	28	**c**
5	**c**	13	**a**	21	**b**	29	**b**
6	**c**	14	**d**	22	**d**	30	**b**
7	**b**	15	**c**	23	**a**	31	**b**
8	**d**	16	**d**	24	**a**		

@@@@@@@@@@

6. Introduction to Digital Electronics

RATIONALE

Knowledge of digital electronics is essential as it is widely used in various applications such as communications, embedded systems, computers, consumer and industrial electronics, and military equipment. Digital systems have the advantages like ease of design, higher accuracy, programmability, and noise immunity. Hence, it becomes essential to study and understand the fundamentals of digital electronic devices.

UNIT OUTCOMES

U6-O1: Unit-6 Learning Outcome-1

To know about the basic number systems and their interconversions

U6-O2: Unit-6 Learning Outcome-2

To know about the basic building blocks of a digital logic circuit.

U6-O3: Unit-6 Learning Outcome-3

To study the different binary codes and Boolean algebra rules

U6-O3: Unit-6 Learning Outcome-4

To understand the basics of digital circuits such as combinational and sequential circuits.

LEARNING OBJECTIVES

LO1: To understand the number systems.

LO2: To study the number representation in the binary system.

LO3: To study the interconversions in the number system.

LO4: To study the operations of logic gates

LO5: To study the concept of the Universal gates

LO6: To study the Boolean algebra

LO7: To study the combinational and sequential circuit basics.

MAPPING THE UNIT OUTCOMES WITH THE COURSE OUTCOMES

Unit Outcome	EXPECTED MAPPING WITH COURSE OUTCOMES (1- Weak Correlation; 2- Medium correlation; 3- Strong Correlation)						
	CO-1	**CO-2**	**CO-3**	**CO-4**	**CO-5**	**CO-6**	**CO-7**
U3-O1	**1**	--	--	--	--	--	**3**
U3-O2	**1**	--	--	--	--	--	**3**
U3-O3	**1**	--	--	--	--	--	**3**

Interesting Facts:

1. The Indian scholar ***Pingala*** (c. 2nd century BC) developed a binary system for describing prosody. He used binary numbers in the form of short and long syllables (the latter equal in length to two short syllables). They were known as *laghu* (light) and *guru* (heavy) syllables.
2. Boolean algebra was introduced by ***George Boole*** in his first book "The Mathematical Analysis of Logic" in 1847 and set forth more fully in his "An Investigation of the Laws of Thought" in 1854.
3. Very-Large Scale Integration (VLSI) consists of thousands of gates or transistors and performs computational operations such as processors, large memory arrays, and programmable logic devices.

Video Resources:

Sr.	Title	URL	QR Code
1.	Introduction to Digital Signal	https://www.youtube.com/watch?v=M0mx8S05v60&list=PLBlnK6fEyqRjMH3mWf6kwqiTbT798eAOm	
2.	Introduction to the number systems	https://www.youtube.com/watch?v=crSGS1uBSNQ	
3.	Introduction to Boolean Algebra (Part 1 and 2)	https://www.youtube.com/watch?v=WW-NPtlzHwk&list=PLBlnK6fEyqRjMH3mWf6kwqiTbT798eAOm&index=8 https://www.youtube.com/watch?v=OjWmVCG8PLA&list=PLBlnK6fEyqRjMH3mWf6kwqiTbT798eAOm&index=9	
4.	Understanding logic gates	https://www.youtube.com/watch?v=INEtYZqtjTo	
5.	The Universal Gates	https://www.youtube.com/watch?v=cNFgilYDxuA	

6.1. The number system

The study of number systems is essential to understand the representation of data before it is processed in any digital system. The numerical values of any physical quantities can be represented in two different formats i.e. analog and digital. In analog format, the numerical value of any quantity is represented as a continuous range of values between the two expected extreme values. On the other hand, in digital format, the numerical value of the quantity is represented in the steps of discrete values i.e. mostly represented using binary format.

6.1.1. Decimal Number System

The decimal number system is a radix (base) 10 number system as the total number of digits available in the decimal number system is 10 (0 to 9). It doesn't imply that these 10 digits will represent only 10 quantities, but all the other digits can be expressed with the help of these numbers. The place values or weights of different digits in a decimal number system, starting from the decimal point, are 10^0, 10^1, $10^{2,}$ and so on for the integer part and 10^{-1}, 10^{-2}, $10^{-3,}$ and so on for the fractional part. The value or magnitude of a given decimal number can be expressed as the sum of the various digits multiplied by their place values or weights.

For example, for a decimal number 7791.917, the integer part can be represented as:

$$\mathbf{7791 = 7 \times 10^3 + 7 \times 10^2 + 9 \times 10^1 + 1 \times 10^0 = 7000 + 700 + 90 + 1 = 7791}$$

and the fractional part can be represented as:

$$\mathbf{917 = 9 \times 10^{-1} + 1 \times 10^{-2} + 7 \times 10^{-3} = 0.9 + 0.01 + 0.007 = 0.917}$$

It is seen that the place value of any digit is a function of the radix of the concerned number system and the position of the digits. This concept is equally valid for the other number systems also.

Complements are used in digital circuits because it is faster to subtract by adding complements than by performing true subtraction. In the decimal number system, the 9's and 10's complements can be calculated.

The 9's complement of a number can be obtained by subtracting every digit of a number by 9. Consider the numbers such as 6, 27, 234, 672 and the 9's complement of these numbers can be obtained as:-

9-6 = 3, 99-27 = 72, 999-234 = 765, 999-672 = 327, thus 3, 72, 234, 672 are the 9's complement of the numbers described above.

Significance of 9's Complement: The procedure of subtraction can be done in much easier steps with the aid of 9's complement. When we subtract two numbers then we subtract the subtrahend from the minuend. However, with 9's complement, we need not do subtraction. In this procedure, we just need to add 9's complement of the subtrahend to the minuend.

When we subtract the smaller number from the larger one, the 9's complement of subtrahend when added with minuend will result in the formation of carry. We need to add this carry to the result. The resultant addition will be the final answer.

When we subtract a larger number from the smaller one, the result of the addition of the subtrahend with minuend will not produce any carry. When the addition of 9's complement doesn't generate any carry it indicates that the resultant product is negative. The final answer can be obtained by again taking the 9's complement of this number.

When a smaller number is to be subtracted from a larger one

Regular Subtraction	Subtraction using 9's complement
678	678
-234	+765 (9s complement of 234)
—	—
444	① 443
	+1
	—
	444

When a larger number is to be subtracted from a smaller one

Regular Subtraction	Subtraction using 9's complement
228	228
-485	+514 (9's complement of 485)
—	—
-257	742 (No carry indicates –ve value)
	-257(9's complement of the result)

Fig. 6.1: Illustration of 9's complement arithmetic

The 10's complement of any decimal number can be obtained by adding 1 to the 9's complement of the same number.

Consider some decimal numbers 7, 34, 566, and 3456. Let's find 9's complement of each of these.

9-7 = 2, 99-34 = 65, 999-566 = 433, 9999-3456 = 6543

Here, 2, 65, 433 & 6543 are the 9's complement of the above-considered numbers. Now, to find 10's complement of these numbers let's add 1 to each of these numbers.

2+1 = 3, 65+1 = 66, 433+1 = 434, 6543+1 = 6544

Thus, 3, 66, 434 & 6544 are the 10's complement of the numbers 7, 34, 566, and 3456 respectively.

Significance of 10's complement: The 10's complement of the decimal number is crucial in subtracting one number from another. The subtraction can be achieved with the help of the addition of 10's complement of subtrahend with minuend.

(i) When a smaller number is to be subtracted from the larger number, a carry will be generated. Ignore this carry and the rest of the digits of the addition will be the answer.

(ii) When a larger number is to be subtracted from a smaller number, the answer will be negative. There will be no carry generation after the addition of the 10's complement of subtrahend with minuend. This indicates that the resultant answer is negative. The final answer can be evaluated by taking the 10's complement of the number which is obtained after the addition of 10's complement of subtrahend with minuend.

When a smaller number is to be subtracted from a larger one

Regular Subtraction	Subtraction using 10's complement
678	678
-234	+766 (10's complement of 234)
------	------
444	① 444 (Ignore the carry)

	444

When a larger number is to be subtracted from a smaller one

Regular Subtraction	Subtraction using 10's complement
228	228
-485	+515 (10's complement of 485)
------	------
-257	743 (No carry indicates –ve value) -257(10's complement of the result)

Fig. 6.2: Illustration of 10's complement arithmetic

6.1.2. Binary Number System

A binary number is a number expressed in the base-2 numeral system or binary numeral system, a method of mathematical expression that uses only two symbols: typically "0" (zero) and "1" (one). The base-2 numeral system is a positional notation with a radix of 2. Starting from the binary point, the place values of different digits in a mixed binary number are 2^0, 2^1, 2^2, and so on, for the integer part and 2^{-1}, 2^{-2}, 2^{-3}, and so on for the fractional part.

Here, each digit is referred to as a bit or binary digit. A byte is a string of eight bits. The byte is the basic unit of data operated in computers. The word length or word size may vary from computer to computer. The word length may equal one byte, two bytes, four bytes, or be even larger.

1's complement of a binary number is another binary number obtained by toggling all bits in it, i.e., transforming the 0 bit to 1 and the 1 bit to 0.

Examples:

1's complement of "0111" is "1000"

1's complement of "1100" is "0011"

2's complement of a binary number is 1, added to the 1's complement of the binary number. In the 2's complement representation of binary numbers, the MSB represents the sign with a '0' used for plus sign and a '1' used for a minus sign. the remaining bits are used for representing magnitude. positive magnitudes are represented in the same way as in the case of sign-bit or 1's complement representation. Negative magnitudes are represented by the 2's complement of their positive counterparts.

Examples:

2's complement of "0111" is "1001"

2's complement of "1100" is "0100"

6.1.3. Octal Number System

The octal number system has a radix of 8 and therefore has eight distinct digits i.e. 0 to 7. The place values for the different digits in the octal number system are 8^0, 8^1, 8^2, and so on for the integer part and 8^{-1}, 8^{-2},8^{-3}, and so on for the fractional part.

6.1.4 Hexadecimal Number System

The hexadecimal number system has radix-16. Its 16 basic digits are 0 to 15. The digits 10 to 15 can also be represented with letters A, B, C, D, E, and F. Hence, this number system is called an alphanumeric code. The place values or weights of different digits in a mixed hexadecimal number are 16^0, 16^1, 16^2, and so on for the integer part and 16^{-1}, 16^{-2}, 16^{-3}, and so on for the fractional part. The hexadecimal number system provides a compact way of representing large binary numbers to store and process in computers.

6.2 Number Representation in Binary system

The binary representation of a positive and negative decimal number can be carried out by using different methods such as the sign-bit magnitude method, the 1's complement, and the 2's complement method.

6.2.1 Sign-Bit Magnitude

In the sign-bit magnitude representation of positive and negative decimal numbers, the MSB represents the sign-bit. Here, a '0' represents a plus sign, and a '1' represents a minus sign. The remaining bits represent the magnitude. For example, the eight-bit representation of +5 would be 00000101, and −5 would be 10000101. An *n*-bit binary representation can be used to represent decimal numbers in the range of $-(2^{n-1}-1)$ to $+(2^{n-1}-1)$. Hence, an eight-bit representation can be used to represent decimal numbers in the range from −127 to +127 using the sign-bit magnitude format.

6.2.2 1's Complement

1's complement binary numbers are very useful in signed number representation. Positive numbers are simply represented as Binary numbers. On the other hand, if the number is negative, then it is represented using 1's complement. First represent the number with a positive sign and then take 1's complement of that number. For example, the representation of -5 and +5 will be as follows:

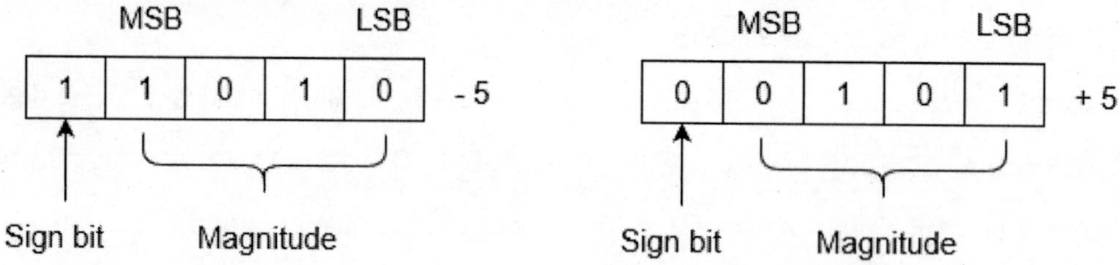

+5 is represented as it is represented in the sign-magnitude method. -5 is represented using the following steps:

(i) +5 = 0 0101

(ii) Take 1's complement of 0 0101 and that is 1 1010. MSB is 1 which indicates that the number is negative. MSB is always 1 in case of negative numbers.

With *n* bits, the range of numbers that can be represented using the 1's complement form will be $-(2^{n-1}-1)$ to $+(2^{n-1}-1)$.

6.2.3 2's Complement

In the 2's complement representation of binary numbers, the MSB represents the sign, with a '0' used for a plus sign and a '1' used for a minus sign. The remaining bits are used for representing magnitude. Positive magnitudes are represented in the same way as in the case of sign-bit or 1's complement representation. Negative magnitudes are represented by the 2's complement of their positive counterparts.

For example, +5 would be represented as 00000101, and −5 would be written as 11111010. The n-bit notation of the 2's complement format can be used to represent all decimal numbers in the range from $+(2^{n-1}-1)$ to $-(2^{n-1})$.

The 2's complement format is very popular due to the ease of generation of 2's complement of a binary number. Further, arithmetic operations are relatively easy to perform when the numbers are represented in the 2's complement format.

6.3. The Decimal Equivalent

The decimal equivalent of a given number in another number system is given by the sum of all the digits multiplied by their respective weights or place values. The integer and fractional parts of the given number should be treated separately.

6.3.1 Binary-to-Decimal Conversion

The decimal equivalent of the binary number $(0101.1001)_2$ is calculated as follows:

- The integer part = 0101
- The decimal equivalent = $0 \times 2^3 + 1 \times 2^2 + 0 \times 2^1 + 1 \times 2^0 = 0 + 4 + 0 + 1 = 5$
- The fractional part = .1001
- Therefore, the decimal equivalent = $1 \times 2^{-1} + 0 \times 2^{-2} + 0 \times 2^{-3} + 1 \times 2^{-4} = 0.5 + 0 + 0 + 0.0625 = 0.5625$
- Therefore, the decimal equivalent of $(0101.1001)_2$ =$(5.5625)_{10}$

6.3.2 Octal-to-Decimal Conversion

The decimal equivalent of the octal number is determined as follows:

- The integer part = 517
- The decimal equivalent = $5 \times 8^2 + 1 \times 8^1 + 7 \times 8^0 = 320 + 8 + 7 = 335$
- The fractional part = .09
- The decimal equivalent = $0 \times 8^{-1} + 9 \times 8^{-2} = 0.1406$
- Therefore, the decimal equivalent of $(517.09)_8 = (335.1406)_{10}$

6.3.3 Hexadecimal-to-Decimal Conversion

The decimal equivalent of the hexadecimal number $(9B1.20)_{16}$ is calculated as-

- The integer part = 9B1
- The decimal equivalent = $9 \times 16^2 + 11 \times 16^1 + 1 \times 16^0 = 2304 + 176 + 1 = 2481$
- The fractional part = 20
- The decimal equivalent = $2 \times 16^{-1} + 0 \times 16^{-2} = 0.125+0=0.125$
- Therefore, the decimal equivalent of $(9B1.20)_{16} = (2481.125)_{10}$

6.3.4 Decimal-to-Binary Conversion

Here, the integer and fractional parts are separately operated. For the integer part, the binary equivalent can be found by successively dividing the number by 2 and recording the remainder until the quotient becomes 0. The remainders written in reverse order constitute the binary equivalent for a given number. The fractional part can be calculated by successively multiplying the fractional part of the decimal number by 2 and recording the carry until the result of the multiplication is ‘0’. The carry sequence written in forward order constitutes the binary equivalent of the fractional part of the decimal number.

In the case of the fractional part, if the result of multiplication does not seem to be converging to zero, the process may be continued only until the requisite number of equivalent bits has been obtained.

Ex. 6.1: Find the binary equivalent of $(17.975)_{10}$.

Solution: The integer part = 13

Divisor	Dividend	Remainder
2	17	--
2	8	1
2	4	0
2	2	0
2	1	0
--	0	1

- The binary equivalent of $(17)_{10}$ is therefore $(10001)_2$
- The fractional part = .975
- 0.975 × 2 = 0.95 with a carry of 1
- 0.95 × 2 = 0.9 with a carry of 1
- 0.9 × 2 = 0.8 with a carry of 1
- The binary equivalent of $(0.975)_{10} = (.111)_2$
- Therefore, the binary equivalent of $(17.975)_{10} = (10001.111)_2$

6.3.5 Decimal-to-Octal Conversion

Here, the progressive division in the case of the integer part and the progressive multiplication while working on the fractional part are carried out by the radix of the octal number system i.e.8. The integer and fractional parts of the decimal number are treated separately.

Ex. 6.2: Find the octal equivalent of $(72.17)_{10}$

Solution: The integer part = 73

Divisor	Dividend	Remainder
8	72	--
8	9	0
8	1	1
--	0	1

- The octal equivalent of $(72)_{10} = (110)_8$
- The fractional part = 0.17
- 0.17 × 8 = 0.36 with a carry of 1
- 0.36×8=0.88 with a carry of 2
- 0.88 × 8=0.04 with a carry of 7
- The octal equivalent of $(0.17)_{10} = (.127)_8$
- Therefore, the octal equivalent of $(72.17)_{10} = (110.127)_8$

6.3.6 Decimal-to-Hexadecimal Conversion

Here, the progressive division in the case of the integer part and the progressive multiplication while working on the fractional part are carried out by the radix of the octal number system i.e.16. The integer and fractional parts of the decimal number are treated separately.

Ex. 6.3: Determine the hexadecimal equivalent of $(28.25)_{10}$

Solution: The integer part = 28

Divisor	Dividend	Remainder
16	28	--
16	1	12=C
--	0	1

- The hexadecimal equivalent of $(28)_{10} = (1C)_{16}$
- The fractional part = 0.25

- $0.25 \times 16 = 0$ with a carry of 4
- Therefore, the hexadecimal equivalent of $(28.25)_{10} = (1C.4)_{16}$

6.3.7 Binary–Octal and Octal–Binary Conversions

The octal number system provides a convenient way of converting large binary numbers into more compact and smaller groups. A binary number can be converted into an octal number by using direct or indirect methods. With the Indirect method, a binary number needs to be converted into another base system (e.g., into decimal, or hexadecimal) and it is to be converted into its equivalent octal number.

Most Significant Bit (MSB)	Octal Point		Least Significant Bit (LSB)		
8^2	8^1	8^0	8^{-1}	8^{-2}	8^{-3}
64	8	1	$\frac{1}{8}$	$\frac{1}{64}$	$\frac{1}{512}$

The numbers are the type of positional number system. It means the weight of the positions from right to left are as 8^0, 8^1, 8^2, 8^3 *and so on* for the integer part and weight of the positions from left to right are 8^{-1}, 8^{-2}, 8^{-3}*and so on* for the fractional part.

Ex. 6.4. Convert the binary number 10010110 into an octal number.

Solution: First convert this into a decimal number

$= (10010110)_2$

$= 1x2^7+0x2^6+0x2^5+1x2^4+0x2^3+1x2^2+1x2^1+0x2^0$

$= 128+0+0+16+0+4+2+0$

$= (150)_{10}$

Then, convert it into octal number

Divisor	Dividend	Remainder
8	150	--
8	18	6
8	2	2
8	0	2

$= (226)_8$

However, there is also a direct method to convert a binary number into an octal number, known as a grouping.

Using Grouping

There are only 8 digits (from 0 to 7) in an octal number system. Hence, we can represent any digit of the octal number system using only 3 bits as given in Table 6.1 below.

Table 6.1: Octal to Binary Conversion

Octal Digit Value	Binary Equivalent
0	000
1	001
2	010
3	011
4	100
5	101
6	110
7	111

Hence, if we make each group of 3 bits of binary input number, then replace each group of binary number with its equivalent octal digits. That will be an octal number of the given number. Note that you can add any number of 0's in the leftmost bit (or in the most significant bit) for the integer part and add any number of 0's in the rightmost bit (or in the least significant bit) for the fraction part for completing the group of 3 bit, this does not change the value of input binary number.

Following are the steps to convert a binary number into an octal number-

- Take binary number
- Divide the binary digits into groups of three (starting from the right) for the integer part and start from the left for the fraction part.
- Convert each group of three binary digits to one octal digit.

Ex. 6.5. Convert binary number 1010111100 into an octal number.

Solution: Since there is no binary point here and no fractional part.

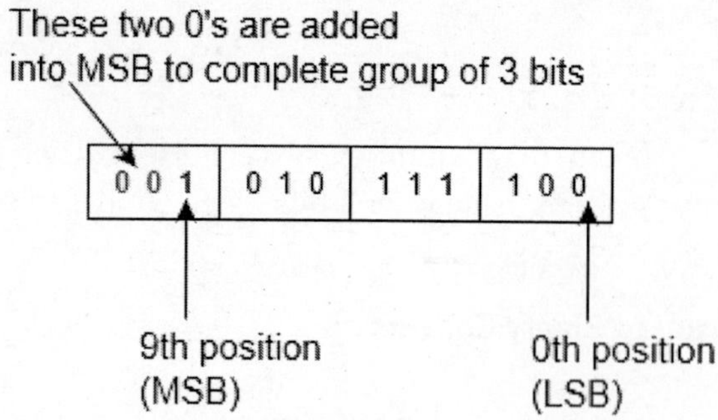

Therefore, Binary to octal is.

$= (1010111100)_2$

$= (\underline{001}\ \underline{010}\ \underline{111}\ \underline{100})_2$

$= (1\ 2\ 7\ 4)_8$

$= (1274)_8$

Ex. 6.6. Convert binary number 0110 011.1011 into octal number.

Solution: Since there is the binary point here and a fractional part. Hence,

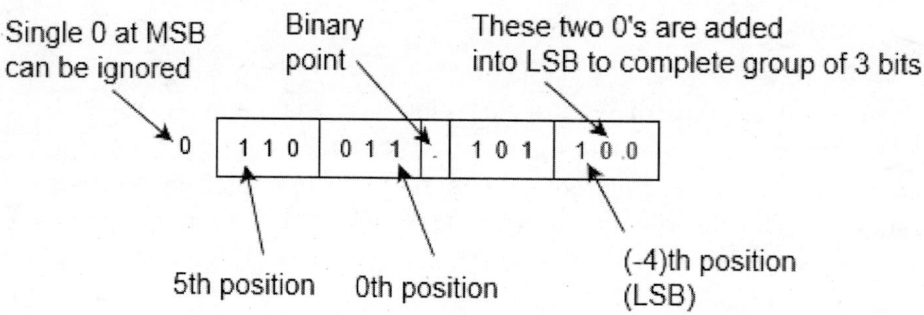

Therefore, Binary to octal is-

$= (0110\ 011.1011)_2$

$= (0\ 110\ 011\ .\ 101\ 1)_2$

$= (110\ 011\ .\ 101\ 100)_2$

$= (6\ 3\ .\ 5\ 4)_8$

$= (63.54)_8$

Octal to Binary Number Conversion:

There are various direct or indirect methods to convert an octal number into a binary number. In an indirect method, one needs to convert an octal number into another number system (e.g., decimal or hexadecimal), then it can be converted into a binary number by converting each digit into a binary number from the hexadecimal system and using a conversion system from decimal to binary number.

Ex. 6.7. Convert the octal number 205 into a binary number.

Solution: First convert it into a decimal or hexadecimal number,

$= (205)_8 = (2x8^2+0x8^1+5x8^0)_8$ *or* $(010\ 000\ 101)_2$

Because the base of octal and hexadecimal are 8 and 16 respectively.

$= (133)_{10}$ *or* $(\underline{0}\ \underline{1000}\ \underline{0101})_2$

$= (133)_{10}$ or $(85)_{16}$

Then convert it into a binary number by converting each digit.

$= (1x2^7+0x2^6+0x2^5+0x2^4+0x2^3+1x2^2+0x2^1+1x2^0)_2$ *or* $(1000\ 0101)_2$

$= (10000101)_2$

However, there is a simple direct method to convert an octal number to a binary number. Since there are only 8 symbols (i.e., 0, 1, 2, 3, 4, 5, 6, and 7) in the octal representation system and its base (i.e., 8) is equivalent to 2^3=8. Hence, we can represent each digit of an octal in a group of 3 bits in the binary number.

Octal Symbol	Binary equivalent
0	000
1	001
2	010
3	011
4	100
5	101
6	110
7	111

This method is simple and also works as the reverse of Binary to Octal Conversion. The algorithm is explained below.

- Take the Octal number as input
- Convert each digit of an octal into binary.
- That will be output as a binary number.

Ex.6.8. Convert the octal number 540 into a binary number.

Solution: According to the above algorithm, the equivalent binary number will be,

$= (540)_8$

$= (101\ 100\ 000)_2$

$= (101100000)_2$

Ex. 6.9. Convert the octal number 352.563 into a binary number.

Solution: According to the above algorithm, the equivalent binary number will be,

$= (352.563)_8$

$= (011\ 101\ 010\ .\ 101\ 110\ 011)_2$

$= (011101010.101110011)_2$

6.3.8 Binary–Hex and Hex–Binary Conversions

The hexadecimal number system provides a convenient way of converting large binary numbers into more compact and smaller groups.

There are various ways to convert a binary number into a hexadecimal number. You can convert using direct methods or indirect methods. First, you need to convert a binary into another base system (e.g., into decimal, or octal). Then you need to convert its hexadecimal number.

Most Significant Bit (MSB)	Hexadecimal Point		Least Significant Bit (LSB)		
16^2	16^1	16^0	16^{-1}	16^{-2}	16^{-3}

Most Significant Bit (MSB)	Hexadecimal Point		Least Significant Bit (LSB)		
256	16	1	$\frac{1}{16}$	$\frac{1}{256}$	$\frac{1}{4096}$

Since numbers are a type of positional number system. That means the weight of the positions from right to left are as 16^0, 16^1, 16^2, 16^3, and so on. For the integer part and weight of the positions from left to right are as 16^{-1}, 16^{-2}, 16^{-3}, and so on for the fractional part.

Ex. 6.10. Convert binary number 1101010 into a hexadecimal number.

Solution: First convert the given number into a decimal number:

$= (1101010)_2$

$= 1x2^6+1x2^5+0x2^4+1x2^3+0x2^2+1x2^1+0x2^0$

$= 64+32+0+8+0+2+0$

$= (106)_{10}$

Then, convert it into a hexadecimal number

$= (106)_{10}$

$= 6x16^1+10x16^0$

$= (6A)_{16}$ is the answer.

However, there is also a direct method to convert a binary number into a hexadecimal number – a grouping that is explained below.

Using Grouping

Since there are only 16 digits (from 0 to 7 and A to F) in the hexadecimal number system, we can represent any digit of the hexadecimal number system using only 4 bits as follows below.

Hexadecimal	0	1	2	3	4	5	6	7
Binary	0000	0001	0010	0011	0100	0101	0110	0111
Hexadecimal	8	9	A	B	C	D	E	F
Binary	1000	1001	1010	1011	1100	1101	1110	1111

If you make each group of 4 bits of binary input number, then replace each group of binary number from its equivalent hexadecimal digits. That will be a hexadecimal number of a given number. Note that you can add any number of 0's in the leftmost bit (or in the most significant bit) for the integer part and add any number of 0's in the rightmost bit (or in the least significant bit) for the fraction part for completing the group of 4 bit, this does not change the value of input binary number.

Following are the steps to convert a binary number into a hexadecimal number:

- Take binary number
- Divide the binary digits into groups of four (starting from the right) for the integer part and starting from the left for the fraction part.
- Convert each group of four binary digits to one hexadecimal digit.

Ex. 6.11. Convert binary number 1010101101001 into a hexadecimal number.

Solution: Since there is no binary point here and no fractional part, hence,

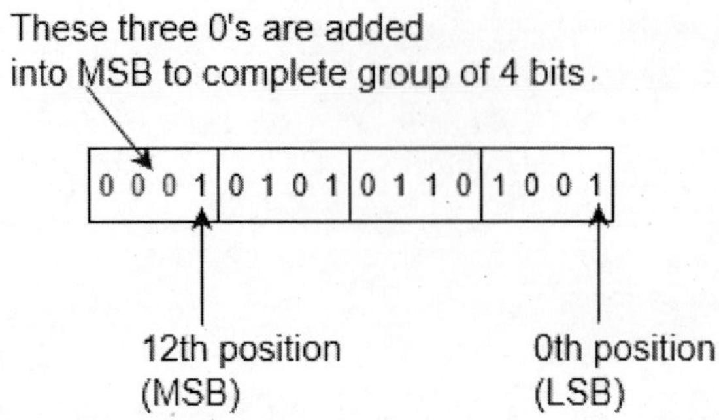

Therefore, Binary to hexadecimal is,

$= (1010101101001)_2$

$= (1\ 0101\ 0110\ 1001)_2$

$= (0001\ 0101\ 0110\ 1001)_2$

$= (1\ 5\ 6\ 9)_{16}$

$= (1569)_{16}$

Ex.6.12. Convert binary number 001100101.110111 into a hexadecimal number.

Solution: The example consists of a binary point here and a fractional part. Hence,

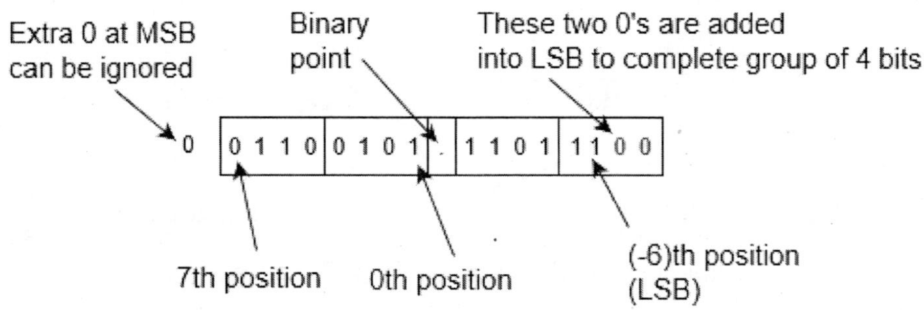

Therefore, Binary to hexadecimal is,

= $(001100101.110111)_2$

= $(0\ 0110\ 0101\ .\ 1101\ 1100)_2$

= $(0110\ 0101\ .\ 1101\ 1100)_2$

= $(6\ 5\ .\ D\ C)_{16}$

= $(65.DC)_{16}$

Hexadecimal to binary conversion is done to obtain the equivalent binary number of the hexadecimal. Converting hexadecimal to binary cannot be done directly. The hexadecimal number has to be converted to a decimal number and then converted to a binary number. One of the most important aspects to remember here is every hexadecimal number will produce 4 binary digits. The hexadecimal to binary conversion can be carried out in two ways, i.e. indirect and direct methods. In the indirect method, first, the hexadecimal is converted to a decimal number and then by using the division process the equivalent binary number can be obtained. In the direct method, we can use the hexadecimal to decimal to the binary conversion table. Let us look at the steps of both methods.

Steps for the indirect method:

This method requires both multiplication and division of numbers using the respective base numbers. The hexadecimal base number is 16, the base number of a decimal number is 10, and the base of a binary number is 2. Let us look at the steps:

- Step 1: Write the hexadecimal number and find its equivalent decimal number.

➢ Step 2: To find the decimal equivalent, multiply each digit with 16^{n-1}, where the digit is in its nth position.

➢ Step 3: After multiplying the numbers, add the product of those numbers to obtain the decimal number.

➢ Step 4: To convert decimal to binary, divide the decimal number by 2 and keep the remainders aside, and divide the quotient by 2 until we arrive at zero.

➢ Step 5: Once the quotient is zero, arrange the remainder from bottom to top i.e. reverse order to obtain the binary number.

Ex. 6.13: Convert hexadecimal $(100)_{16}$ to binary.

Solution:

$(100)_{16} = 1 \times 16^2 + 0 \times 16^1 + 0 \times 16^0$

=1x256+0+0 =256

Therefore, (100)16 = (256)10

Now convert, (256)10 to its binary equivalent as-

2	256	 0
2	128	 0
2	64	 0
2	32	 0
2	16	 0
2	8	 0
2	4	 0
2	2	 0
	1	

$\therefore 256_{10} = 100000000_2$

Hence, $(100)_{16} = (100000000)_2$

In the direct method, just looking at the conversation table we can convert hexadecimal to binary. Table 6.2 shows the hexadecimal to the binary conversion.

Table 6.2: Hexadecimal to Binary Conversion

Hexadecimal Digit	Decimal Digit	Binary Digit
0	0	0000

1	1	0001
2	2	0010
3	3	0011
4	4	0100
5	5	0101
6	6	0110
7	7	0111
8	8	1000
9	9	1001
A	10	1010
B	11	1011
C	12	1100
D	13	1101
E	14	1110
F	15	1111

6.3.9. Hex–Octal and Octal–Hex Conversions

For hexadecimal–octal conversion, the given hex number is firstly converted into its binary equivalent which is further converted into its octal equivalent. An alternative approach is first to convert the given hexadecimal number into its decimal equivalent and then convert the decimal number into an equivalent octal number. For octal–hexadecimal conversion, the octal number may first be converted into an equivalent binary number and then the binary number is transformed into its hex equivalent. The other option is first to convert the given octal number into its decimal equivalent and then convert the decimal number into its hex equivalent.

Ex. 6.14. Convert $(2CD)_{16}$ to an Octal number

Solution: The given hex number = $(2CD)_{16}$

2 → 0010, C →1100, D → 1101
Now you will be grouping them from right to left, each having 3 digits.

001, 011, 001, 101

001→1, 011 →3, 001→1, 101→5

Hence, $(2CD)_{16} = (1315)_8$

Ex. 6.15. Convert $(536)_8$ to a hexadecimal number

Solution: Converting $(536)_8$ into its binary equivalent we get

5→101, 3→011,6→110

$=(101011110)_2$

Now forming the group of 4 binary bits to obtain its hexadecimal equivalent,

$(101011110)_2$= (0001) (0101) (1110)

= $(15E)_{16}$

Hence, the equivalent hexadecimal number of 536 is 15E.

6.4. Binary Codes

6.4.1 Binary Coded Decimal (BCD)

The binary-coded decimal (BCD) code is used to represent a given decimal number in its equivalent binary form. The BCD equivalent of a decimal number can be written by replacing each decimal digit in the integer and fractional parts with its four-bit binary equivalent.

As an example, the BCD equivalent of $(09.17)_{10}$ is written as $(0000\ 1001.0001\ 0111)_{BCD}$. This representation of BCD code is known as the 8421 BCD code; 8, 4, 2, and 1 represent the weights of different bits in the four-bit groups, starting from MSB and proceeding towards LSB. The other weighted BCD codes include the 4221 BCD and 5421 BCD codes; with each bit in the code representing the weights of the relevant bits. Table 6.3 shows the 8421, 4221, and 5421 BCD codes.

Table 6.3: BCD code representations

Decimal	8421 BCD code	4221 BCD codes	5421 BCD codes
0	0000	0000	0000
1	0001	0001	0001

2	0010	0010	0010
3	0011	0011	0011
4	0100	1000	0100
5	0101	0111	1000
6	0110	1100	1001
7	0111	1101	1010
8	1000	1110	1011
9	1001	1111	1100

BCD-to-Binary Conversion:

A given BCD number can be converted into an equivalent binary number by first writing its decimal equivalent and then converting it into its binary equivalent.

Ex. 6.16. Convert $(0001\ 0111.0000\ 0111)_{BCD}$ to its equivalent binary number.

Solution:

- Corresponding decimal number: 17.07
- The binary equivalent of 17.07 can be determined to be 10001 for the integer part and 0.0010 for the fractional part.
- Therefore, $(0001\ 0111.0000\ 0111)_{BCD} = (10001.0010)_2$.

Binary-to-BCD Conversion:

The process of binary-to-BCD conversion is the same as the process of BCD-to-binary conversion performed in the opposite order. A given binary number can be converted into an equivalent BCD number by calculating its decimal equivalent and then writing the corresponding BCD equivalent.

Ex. 6.17. Convert $(10001111011000.0001)_2$ to its BCD equivalent number.

Solution:

- The decimal equivalent of this binary number is 9176.07
- Using the decimal equivalent, the BCD equivalent can then be written as- 1001 0001 0111 0110.0000 0111

6.4.2. Excess-3 Code

The excess-3 code is also treated as XS-3 code. The excess-3 code is a non-weighted and self-complementary BCD code used to represent decimal numbers. This code plays an important role in arithmetic operations as it resolves deficiencies encountered when we use the 8421 BCD code for adding two decimal digits whose sum is greater than 9. The Excess-3 code uses a special type of algorithm, which differs from the binary positional number system or normal non-biased BCD. We can find the excess-3 code of the given binary number by using the following steps:

1. Find the decimal number of the given binary number.
2. Add 3 in each digit of the decimal number.
3. Find the 4-bit binary code of each digit of the newly generated decimal number.

It is to be noted that, if the addition of 3 to a digit produces a carry (it may be in the case with the digits 7, 8, and 9); that carry should not be taken forward. The result of addition should be taken as a single entity and subsequently replaced with its excess-3 code equivalent. Table 6.4 lists the excess-3 code for the decimal numbers.

Table. 6.4 Excess-3 code for the decimal numbers

Decimal	Excess-3 code	Decimal	Excess-3 code
0	0011	5	1000
1	0100	6	1001
2	0101	7	1010
3	0110	8	1011
4	0111	9	1100

Ex. 6.18. Calculate the excess-3 code for $(917)_{10}$.

Solution:

- The addition of '3' to each digit gives- 9+3=12, 1+3=4, 7+3=10

- The corresponding four-bit binary equivalents for 12,1 and 10 are 1100, 0100, and 1010 respectively.
- The excess-3 code for $(597)_{10}$ = 1100 0100 1010 = $(110001001010)_{Ex\text{-}3}$

6.4.3. Gray Code

The Gray code is a sequence of binary number systems, which is also known as reflected binary code. The reason for calling this code a reflected binary code is the first $\frac{N}{2}$ values compared with those of the last $\frac{N}{2}$ values in reverse order. In this code, two consecutive values are differed by one bit of binary digits. Gray codes are used in the general sequence of hardware-generated binary numbers. The gray code is a very light weighted code because it doesn't depend on the value of the digit specified by the position. This code is also called a cyclic variable code as the transition of one value to its successive value carries a change of one bit only. Table 6.5 lists the binary and Gray code equivalents of decimal numbers 0–15.

Table 6.5: Gray code

Decimal	Binary code	Gray codes	Decimal	Binary code	Gray codes
0	0000	0000	8	1000	1100
1	0001	0001	9	1001	1101
2	0010	0011	10	1010	1111
3	0011	0010	11	1011	1110
4	0100	0110	12	1100	1010
5	0101	0111	13	1101	1011
6	0110	0101	14	1110	1001
7	0111	0100	15	1111	1000

Binary–Gray Code Conversion:

The following steps are required to convert a binary number into its equivalent Gray code:

- Firstly, record the most significant bit or MSB or the leftmost bit of the given binary data as it is, to have MSB of gray equivalent.
- Now, proceed towards adding the adjacent bits of the binary data starting from MSB with its adjacent bit to LSB. While adding, put the summation obtained in place of the next bit and ignore the carry.
- Repeat the same process for all the bits in the sequence till LSB.

Ex. 6.19. Convert $(110101)_2$ to its equivalent Gray Code.

Solution: The first bit or MSB of the gray equivalent will be the same as the MSB of the binary value. Thus

Binary Code - 1 1 0 1 0 1

Gray Code - 1

Now add the two adjacent bits starting from MSB to LSB and write the result obtained as the next bit.

Binary Code - 1+ 1 0 1 0 1

Gray Code - 1 0

We know the addition of binary 1 and 1 will give 0 as the sum and 1 as the carry. And we have already discussed that the carry bit must be ignored, while the sum bit achieved will be put as the next bit in the gray value. Further, by repeating the same process,

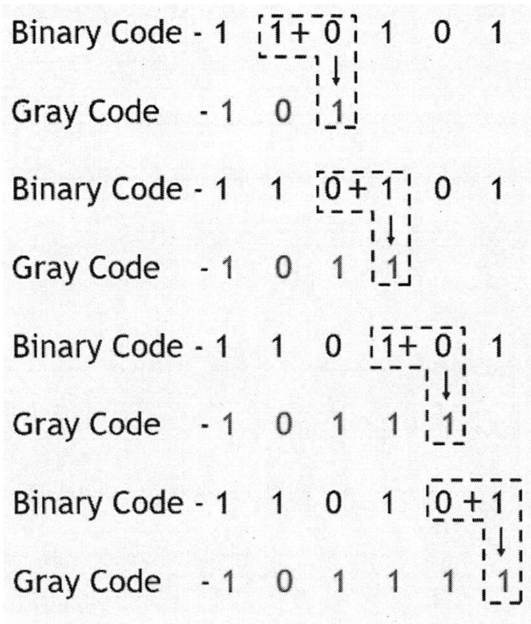

Hence, $(110101)_2$ = $(101111)_{Gray}$

Gray Code–Binary Conversion:

The following steps are required to convert a Gray code to its equivalent binary number:

- Like in the case of binary to gray conversion, here also while writing binary code from gray code, the MSB must remain the same. Hence, write the leftmost bit of gray code as the MSB of binary code.
- Now, add the recently achieved binary digit with the next adjacent gray code bit. The sum must be written as the next bit of binary equivalent, while the carry must be neglected.
- The above-discussed step must be followed for all the bits present in the sequence.

Ex. 6.20. Convert $(101011)_{Gray}$ to its equivalent binary code.

Solution:

In the first step, write the MSB of gray value as the MSB of binary equivalent.

Gray code - 1 0 1 0 1 1

Binary code - 1

Now on adding the achieved binary bit and the next adjacent gray bit till the LSB of the sequence,

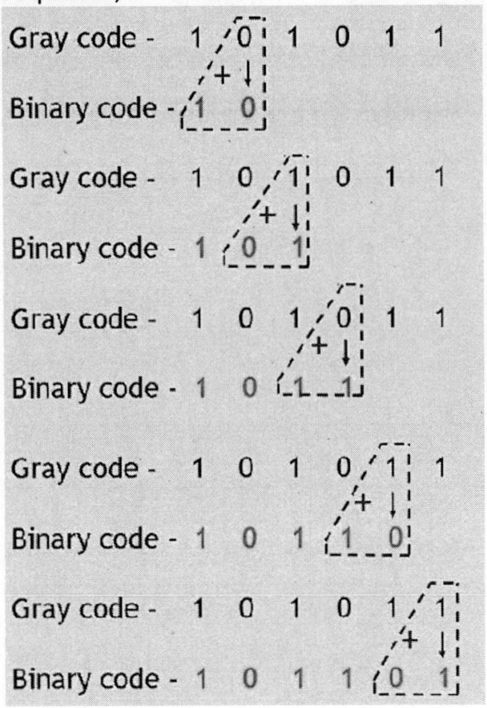

Hence $(101011)_{Gray} = (101101)_2$

6.5. The Digital Arithmetic

The basic rules for binary addition, subtraction, multiplication, and division are stated as follows:

(a) 0 + 0 = 0.

(b) 0 + 1 = 1.

(c) 1 + 0 = 1.

(d) 1 + 1 = 0 with a carry of '1' to the next more significant bit.

The rules for binary subtraction are as follows:

(a) 0 – 0 = 0.

(b) 1 – 0 = 1.

(c) 1 – 1 = 0.

(d) 0 – 1 = 1 with a borrow of 1 from the next more significant bit.

Binary multiplication is the process of repeated addition. The basic rules of multiplication are as follows:

(a) $0 \times 0 = 0$.

(b) $0 \times 1 = 0$.

(c) $1 \times 0 = 0$.

(d) $1 \times 1 = 1$.

Binary Division

Binary division, similar to other binary arithmetic operations, is performed on binary numbers. The algorithm for binary division is somewhat similar to decimal division; the only difference here lies in the rules followed using the digits '0' and '1'. Binary multiplication and binary subtraction are the two binary arithmetic operations that are performed while performing a binary division. The use of only '0' and '1' makes binary division quite simpler in comparison to decimal division. Other operations that are used while performing binary division are binary multiplication and binary subtraction. Following rules must be followed while doing a binary division-

Dividend	Divisor	Result
0	1	0
1	1	1
Division by a 0 is meaningless		

Following are the steps to be followed in a binary division operation:

- **Step 1:** Compare the divisor with the dividend. If the divisor is larger, place 0 as the quotient, then bring the second bit of the dividend down. If the divisor is smaller, multiply it by 1 and the result becomes the subtrahend. Then, subtract the subtrahend from the minuend to get the remainder.
- **Step 2:** Then bring down the next number bit from the dividend portion and perform step 1 again.
- **Step 3:** Repeat the same process until the remainder becomes zero or the whole dividend is divided.

Ex. 6.21. Let A = $(011010)_2$ and B = $(0101)_2$. Perform division of A by B.

Solution: Given Dividend A= $(011010)_2$ and the divisor, B = $(0101)_2$.

Step 1: Since the zero in the most significant bit position doesn't change the value of the number, let's remove it from both the dividend and divisor. So the dividend becomes $(11010)_2$, and the divisor becomes $(101)_2$.

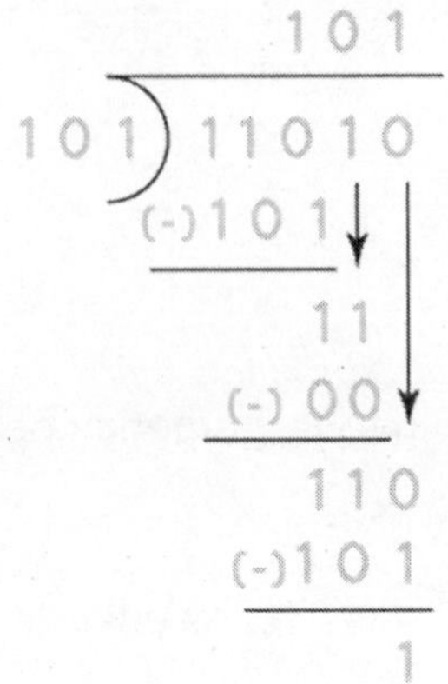

Step 2: Let us use the long-division method. In this step, compare the divisor 101 with the first digit in the dividend 11010, since the divisor is smaller, it will be multiplied by 1 and the result will be the subtrahend. As per the binary multiplication rules: 101 × 1 = 101 and this result is written below.

Step 3: Subtract the subtrahend 101 from the minuend 110. As per the binary subtraction rules: 110 - 101 = 001, this result is written below.

Step 4: As per the rules of division, the next least significant bit comes down, and the divisor is multiplied by 1. Since the result, 101 is bigger than the minuend 0011, this step cannot be completed. Then, we have to go to the next step.

Step 5: We write 0 as the next bit of the quotient and then, the least significant bit 0 comes down.

Step 6: Again the divisor is multiplied by 1 and the result is written as 101 × 1 = 101

Step 7: As per the binary subtraction, subtracting 101 from 110, we get, 110 - 101 = 001. The remainder is similar to Step 3, as all the numbers are the same.

The binary division operation is completed now and we get-

- Quotient = 101
- Remainder = 001 = 1

6.6. The Logic Gates

A logic gate is a device that acts as a building block for digital circuits. They perform basic logical functions that are fundamental to digital circuits. In a circuit, logic gates will make decisions based on a combination of digital signals coming from its inputs. Most logic gates have two (or more) inputs and one output. Logic gates are based on Boolean algebra. At any given moment, every terminal is in one of the two binary conditions, false or true. False represents 0, and true represents 1. Depending on the type of logic gate being used and the combination of inputs, the binary output will differ. There are three basic logic gates, namely the OR gate, the AND gate, and the NOT gate. The NAND gate, the NOR gate, the EXCLUSIVE-OR gate, and the EXCLUSIVE-NOR gate are derived from the basic logic gates.

6.6.1: The OR Gate

An OR gate is a logic circuit with two or more inputs and one output. The OR operation on two independent logic variables A and B is written as Y = A+B and treated as Y equals *A OR B*. The output of an OR gate is LOW only when all of its inputs are LOW and for all other input conditions, the output is HIGH. Fig. 6.3 shows the symbol and the truth table of a two-input OR gate.

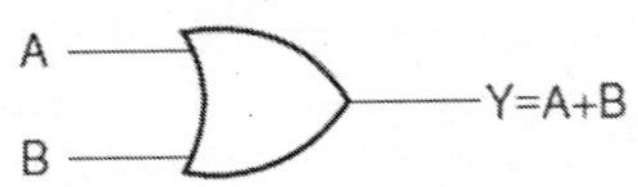

A	B	Y
0	0	0
0	1	1
1	0	1
1	1	1

Fig. 6.3: A Two-input OR gate and its truth table

6.6.2. The AND Gate

An AND gate is a logic circuit having two or more inputs and one output. The AND operation on two independent logic variables A and B is written as Y = A.B and treated as Y equals *A AND B*. The output of an AND gate is HIGH only when all of its inputs are in the HIGH state and all other cases, the output is LOW. Fig. 6.4 depicts the logic symbol and truth table of a two-input AND gate.

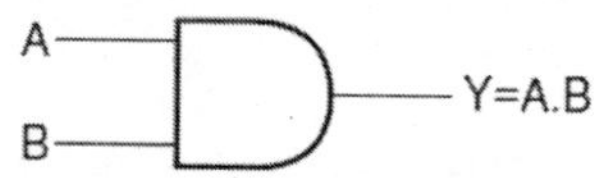

A	B	Y
0	0	0
0	1	0
1	0	0
1	1	1

Fig. 6.4: A Two-input AND gate and its truth table

6.6.3. The NOT Gate

A NOT gate is a one-input, one-output logic circuit whose output is always the complement of the input. A LOW input produces a HIGH output and vice versa. It is also known as an inverting circuit. Fig. 6.5 shows the logic symbol and truth table of a NOT gate. The NOT operation on logic variable X is denoted as $\overline{X}$. Let X be the input to a NOT gate, then its output Y is given by $Y = \overline{X}$ and reads as Y equals NOT X.

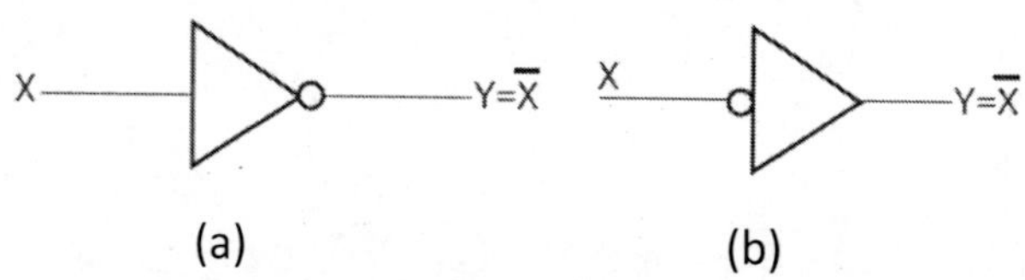

X	Y
0	1
1	0

Fig. 6.5: Symbol of a NOT gate and its truth table

6.6.4. The EXCLUSIVE-OR Gate

The EXCLUSIVE-OR (Ex-OR) gate is a two-input, one-output gate. Fig. 6.6 shows the logic symbol and truth table of a two-input Ex-OR gate. The output of an Ex-OR gate is a logic '1' when the inputs are not similar and a logic '0' when the inputs are similar. The output of a two-input Ex-OR gate is given by -

$$Y = (A \oplus B) = \overline{A}B + A\overline{B}$$

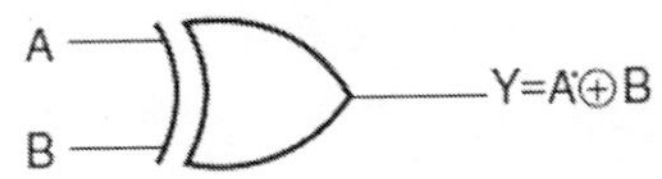

A	B	Y
0	0	0
0	1	1
1	0	1
1	1	0

Fig. 6.6: A two-input Ex-OR gate

6.6.5. The NAND Gate

An AND gate followed by a NOT gate gives rise to a NAND gate. Fig. 6.7(a,b) shows the symbol and the truth table of a two-input NAND gate. The output of a NAND gate is a logic '0' when all its inputs are a logic '1'. For all other input combinations, the output is a logic '1'. The output expression of a two-input NAND gate is given as-$Y = \overline{A.B}$.

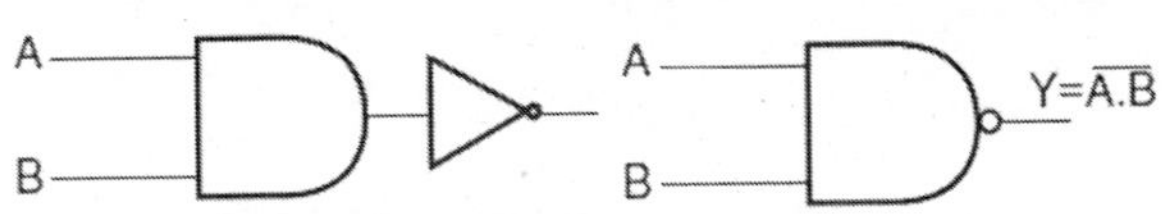

(a) (b)

A	B	Y
0	0	1
0	1	1
1	0	1
1	1	0

Fig. 6.7: A Two input NAND gate and its truth table

6.6.6. The NOR Gate

An OR gate followed by a NOT gate gives rise to a NOR gate. The output of a NOR gate is a logic '1' when all its inputs are logic '0'. For all other input combinations, the output is a logic '0'. The output of a two-input NOR gate is given as $Y = \overline{A + B}$.

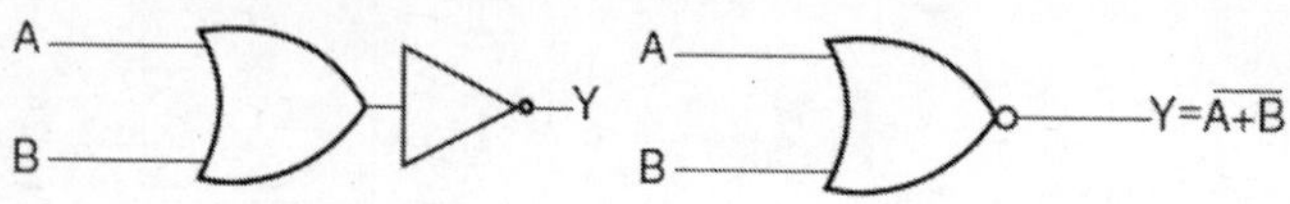

(a) (b)

A	B	Y
0	0	1
0	1	0
1	0	0
1	1	0

Fig. 6.8: A two-input NOR gate and its truth table

6.6.7. The EXCLUSIVE-NOR Gate

An EXCLUSIVE-NOR (Ex-NOR) gate can be realized by complementing the output of an EX-OR gate. Fig 6.9 shows the symbol and the truth table of an Ex-NOR gate. The logic expression for a two-input Ex-NOR gate can be given as $Y = \overline{(A \oplus B)} = \left(A.B + \overline{A}.\overline{B}\right)$

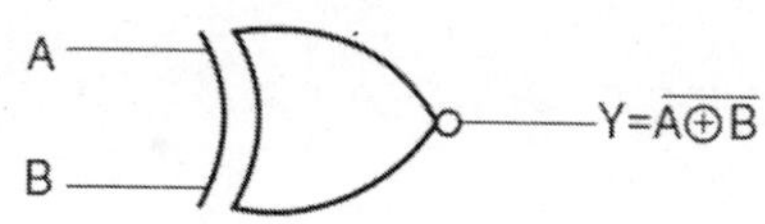

A	B	Y
0	0	1
0	1	0
1	0	0
1	1	1

Fig. 6.9: A two-input Ex-NOR gate and its truth table.

6.6.8. The Universal Gates

A universal gate is a logic gate that can implement any Boolean function without the need to use any other type of logic gate. The NOR gate and NAND gate are universal gates. This means that we can create any logical Boolean expression using only NOR gates or only NAND gates. In practice, this is advantageous since NOR and NAND gates are economical and easier to fabricate than other logic gates. An AND gate is typically implemented as a NAND gate followed by an inverter. Similarly, an OR gate is realized as a NOR gate followed by an inverter. Fig. 6.10 and 6.11 shows NAND and NOR gate as Universal gates, respectively.

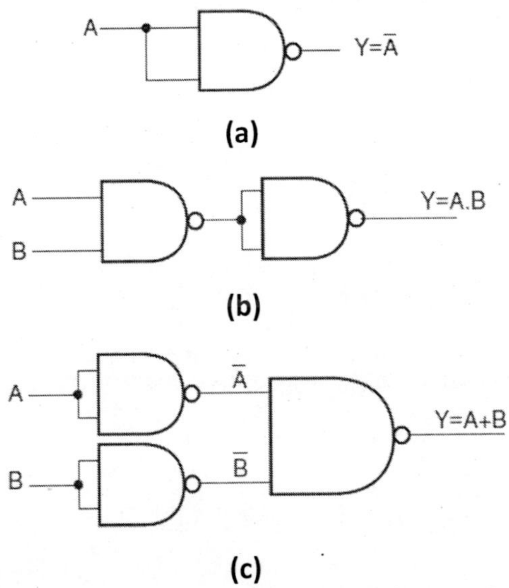

Fig. 6.10: NAND gate as a Universal gate.

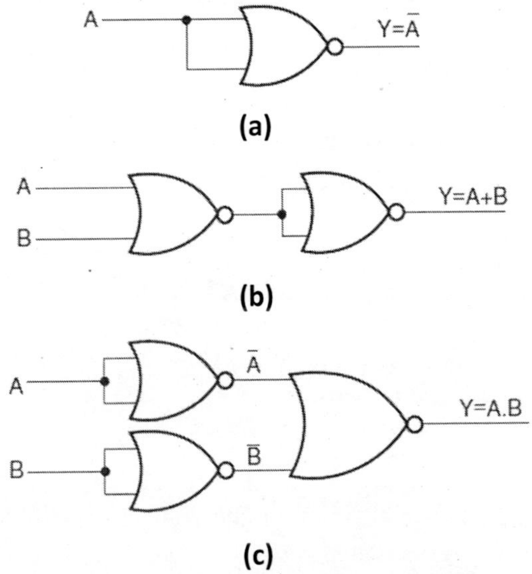

Fig. 6.11: NOR gate as a Universal gate.

6.7. The Boolean algebra

Boolean algebra is a division of mathematics that deals with operations on logical values and incorporates binary variables. Most commonly Boolean variables are presented with the possible values of 1 ("true") or 0 ("false"). Boolean algebra is different from elementary algebra as the latter deals with numerical operations and the former deals with logical operations. Elementary algebra is expressed using basic mathematical functions, such as addition, subtraction, multiplication, and division, whereas Boolean algebra deals with conjunction, disjunction, and negation.

Following are the important rules used in Boolean algebra.

- A variable used can have only two values. Binary 1 for HIGH and Binary 0 for LOW.
- The complement of a variable is represented by an overbar (-). Thus, the complement of variable A is represented as $\overline{A}$. Thus if A = 0 then $\overline{A}$. = 1 and A = 1 then $\overline{A}$. = 0.
- *ORing* of the variables is represented by a plus (+) sign between them. For example, the *ORing* of A, B, and C are represented as A + B + C.
- Logical *ANDing* of the two or more variables is represented by writing a dot between them such as A.B.C. Sometime the dot may be omitted like ABC.

Basic Boolean Laws:

Commutative law

Any binary operation which satisfies the following expression is referred to as a commutative operation.

(i) A.B = B. A (ii) A + B = B + A

Commutative law states that changing the sequence of the variables does not have any effect on the output of a logic circuit.

Associative law

This law states that the order in which the logic operations are performed is irrelevant as their effect is the same.

- A + (B + C) = (A + B) + C = A + B + C (OR Associate Law)
- A(B.C) = (A.B)C = A . B . C (AND Associate Law)
- A + $\bar{A}$B = A + B

Distributive law

Distributive law states the following condition.

- A(B + C) = A.B + A.C (OR Distributive Law)
- A + (B.C) = (A + B).(A + C) (AND Distributive Law)

Absorptive Law

This law enables a reduction in a complicated expression to a simpler one by absorbing like terms.

- A + (A.B) = (A.1) + (A.B) = A(1 + B) = A (OR Absorption Law)
- A(A + B) = (A + 0).(A + B) = A + (0.B) = A (AND Absorption Law)

Annulment Law

A term AND‘ed with a “0” equals 0 or OR‘ed with a “1” will equal 1

- A . 0 = 0   A variable AND’ed with 0 is always equal to 0
- A + 1 = 1   A variable OR’ed with 1 is always equal to 1

Identity Law

A term OR‘ed with a “0” or AND‘ed with a “1” will always equal that term

- A + 0 = A   A variable OR’ed with 0 is always equal to the variable
- A . 1 = A   A variable AND’ed with 1 is always equal to the variable

Idempotent Law

An input that is AND‘ed or OR´ed with itself is equal to that input

- A + A = A   A variable OR’ed with itself is always equal to the variable
- A . A = A   A variable AND’ed with itself is always equal to the variable

Complement Law

A term AND‘ed with its complement equals “0” and a term OR´ed with its complement equals “1”

- A . $\bar{A}$ = 0   A variable AND’ed with its complement is always equal to 0
- A + $\bar{A}$ = 1   A variable OR’ed with its complement is always equal to 1

De-Morgan’s Theorems

(a) First Theorem: The complement of a product of variables is equal to the sum of the complements of the variables : $\overline{\boldsymbol{AB}} = \overline{\boldsymbol{A}} + \overline{\boldsymbol{B}}$

(b) Second Theorem: The complement of a sum of variables is equal to the product of the complements of the variables: $\overline{\mathbf{A}+\mathbf{B}} = \overline{\mathbf{A}}.\overline{\mathbf{B}}$

6.8. The Digital Circuits

Digital circuits are mainly classified into two type’s i.e. combinational and sequential circuits.

6.8.1. The Combinational Circuits

Combinational Logic Circuits are memoryless digital logic circuits whose output at any instant depends only on the combination of its inputs. The outputs of combinational logic circuits are only determined by the logical function of their current input state,. logic "0" or logic "1", at any given instant in time. The result is that combinational logic circuits have no feedback, and any changes to the signals being applied to their inputs will immediately have an effect on the output. Hence, if one of the conditions of its inputs changes state, from 0-1 or 1-0, so too will the resulting output as by default combinational logic circuits have "no memory", "timing" or "feedback loops" within their design. Fig. 6.12 shows a typical combinational logic.

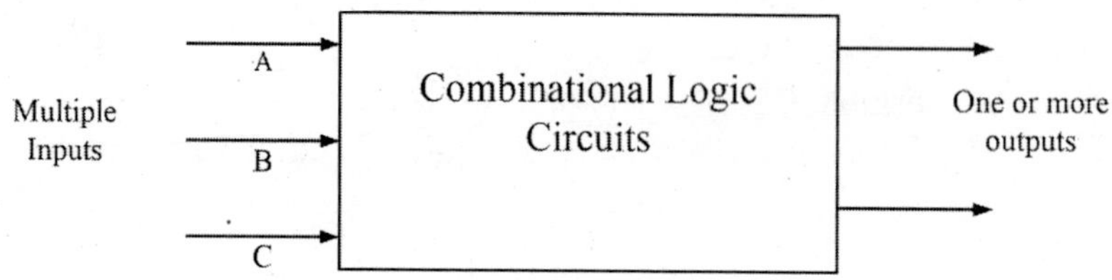

Fig. 6.12. A combinational logic

Combinational logic circuits are made up of basic logic NAND, NOR, or NOT gates that are "combined" or connected to produce more complicated switching circuits. These logic gates are the building blocks of combinational logic circuits. Any combinational circuit can be implemented with only NAND and NOR gates as these are classed as "universal" gates. Common combinational circuits made up of individual logic gates that carry out a desired application include Multiplexers, De-multiplexers, Encoders, Decoders, Full, and Half Adders, etc. The three main ways of specifying the function of a combinational logic circuit are:

1. **Boolean Algebra** – This forms the algebraic expression showing the operation of the logic circuit for each input variable either True or False that results in a logic "1" output.
2. **Truth Table** – A truth table defines the function of a logic gate by providing a concise list that shows all the output states in tabular form for each possible combination of input variables that the gate could encounter.
3. **Logic Diagram** – This is a graphical representation of a logic circuit that shows the wiring and connections of each logic gate, represented by a specific graphical symbol, which implements the logic circuit.

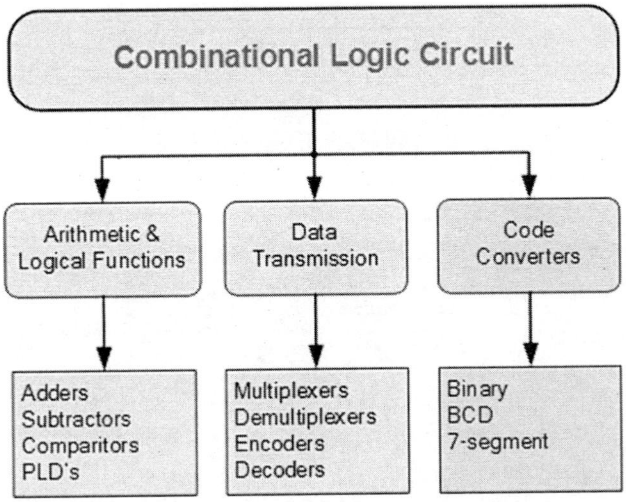

Fig. 6.13: Classification of Combinational Logic

6.8.2. The Sequential Circuits

Unlike combinational logic circuits that change state depending upon the actual signals being applied to their inputs at that time, Sequential Logic circuits have some form of inherent "Memory" built in. The output state of a "sequential logic circuit" is a function of the following three states- the "present input", the "past input" and/or the "past output". Sequential Logic circuits remember these conditions and stay fixed in their current state until the next clock signal changes one of the states, giving sequential logic circuits "Memory".

Sequential logic circuits are generally termed as two-state or bistable devices which can have their output or outputs set in one of two basic states, a logic level "1" or a logic level "0" and will remain "latched" (hence the name latch) indefinitely in this current state or condition until some other input trigger pulse or signal is applied which will cause the bistable to change its state once again. Fig. 6.14 represents a sequential logic. The word "Sequential" means that things happen in a "sequence", one after another, and in Sequential Logic circuits, the actual clock signal determines when things will happen next.

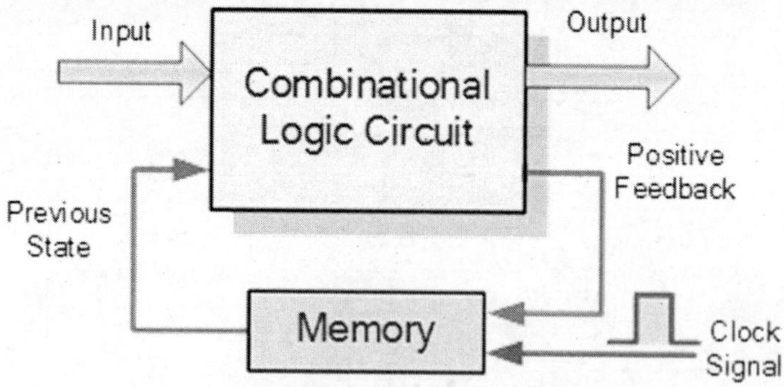

Fig. 6.14: Sequential logic.

Classification of Sequential Logic

As standard logic gates are the building blocks of combinational circuits, bistable latches and flip-flops are the basic building blocks of sequential logic circuits. Sequential logic circuits can be constructed to produce either simple edge-triggered flip-flops or more complex sequential circuits such as storage registers, shift registers, memory devices, or counters. Table 6.6 provides the comparison between sequential and combinational logic circuits. Either way, sequential logic circuits can be divided into the following three main categories:

1. Event Driven – asynchronous circuits that change state immediately when enabled.
2. Clock Driven – synchronous circuits that are synchronized to a specific clock signal.
3. Pulse Driven – which is a combination of the two that responds to triggering pulses.

Table 6.6 : Comparison between the Combinational and Sequential Logic Circuits

Parameters	Combinational Circuit	Sequential Circuit
Working Principle	A Combinational Circuit is a type of circuit in which the output is independent of time and only relies on the input present at that particular instant.	A Sequential circuit is a type of circuit where output not only relies on the current input but also depends on the previous output.
Building blocks	The elementary building blocks of a combinational circuit are its logic gates.	The building blocks of a sequential circuit are the logic gates along with flip flops.
Speed of operation	As the input of current instant is only required in the case of a Combinational circuit, it is faster and better in performance as compared to that of a Sequential circuit.	Sequential circuits are comparatively slower and have low performance as compared to that Combinational circuits.
Availability of the feedback	Since output does not depend on the time instant, no feedback is required for its next output generation.	The output relies on its previous feedback so the output of the previous input is being transferred as feedback used with input for the next output generation.
Ease of implementation	No implementation of feedback makes the combinational circuit less complex as compared to a sequential circuit.	The implementation of feedback makes the sequential circuit more complex as compared to a combinational circuit.
Applications	Combinational circuits are mainly used for arithmetic as well as Boolean operations.	Sequential circuits are mainly used for storing data.

SOLVED EXAMPLES

Ex. 1: Find the decimal equivalent of the following binary numbers expressed in the 2's complement format:

(a) 00001011

(b) 10001011

Solution:

(a) The MSB bit is '0', which indicates a plus sign.

The magnitude bits are 0001011

The decimal equivalent = $0\times2^6 + 0\times2^5 + 0\times2^4 + 1\times2^3 + 0\times2^2 + 1\times2^1 + 1\times2^0$

= 0+0+0+8+0+2+1 = 11

Therefore, 00001011 represents +11

(b) The MSB bit is '1', which indicates a minus sign

The magnitude bits are therefore given by the 2's complement of 0001011, i.e. 1110100

The decimal equivalent = $1\times2^6 + 1\times2^5 + 1\times2^4 + 0\times2^3 + 1\times2^2 + 0\times2^1 + 0\times2^0$

=64+32+16+0+4+0+0=116

Therefore, 10001110 represents 116

Ex.2: Calculate (a) the excess-3 equivalent of $(1707.717)_{10}$ and (b) the decimal equivalent of the excess-3 number 110111101011.00110101.

Solution: (a) For integer part, the Ex-3 number is 1 7 0 7 + 3 3 3 3 = 4 10 3 10

Now replacing digits 4 10 3 10 with the four-bit binary equivalents, we get the excess-3 code for $(1707)_{10}$ as 0100 1010 0011 1010 = 0100101000111010

For fractional part, the excess-3 code for $(.717)_{10}$ is 7+3=10; 1+3=4; 7+3=10. Now these digits with the four-bit binary equivalents as – 1010 0100 1010= 101001001010

Hence the excess-3 code for $(1707.717)_{10}$ =0100101000111010. 101001001010

(b) To obtain the decimal equivalent from the given excess-3 code, subtracting 0011 from each four-bit group-

For integer part-1101-0011=1010;1110-0011=1011;1011-0011=1000

For fractional part-0011-0011=0000; 0101-0011=0010

The decimal equivalent for the integer part=10118

The decimal equivalent for fractional part=02

Hence, the decimal equivalent of $(110111101011.00110101)_{Ex-3}$= $(10118.02)_{10}$.

Ex. 3: Perform the following addition operations:

(a) $(AF1.B3)_{16}$ + $(FFF.E)_{16}$

(b) $(275.75)_{10}$ + $(37.875)_{10}$

Solution:

(a) (AF1.B3)16=$(101011110001.10110011)_2$ and $(FFF.E)_{16}$ =$(111111111111.1110)_2$.

To have the same number of bits in the integer and fractional parts, $(111111111111.1110)_2$ can also be written as $(111111111111.11100000)_2$

The two numbers can now be added as-

0101011110001.10110011

0111111111111.11100000

1101011110001.10010011

The hexadecimal equivalent of $(1101011110001.10010011)_2$ is $(1AF1.93)_{16}$

(b) The given decimal numbers need to be converted into their equivalent binary numbers.

$(275.75)_{10}$ = $(100010011.11)_2$ and $(37.875)_{10}$ = $(100101.111)_2$

The two binary numbers can be rewritten as $(100010011.110)_2$ and $(000100101.111)_2$ to have the same number of bits in their integer and fractional parts. The addition of two numbers is performed as follows:

100010011.110

000100101.111

100111001.101

The decimal equivalent of $(100111001.101)_2$ is $(313.625)_{10}$.

Ex.4: Perform $(185)_{10}$ –$(8)_{10}$ using the excess-3 code.

Solution:

- $(185)_{10}$ = $(0001\ 1000\ 0101)_{BCD}$. The excess-3 equivalent of $(0001\ 1000\ 0101)_{BCD}$ = 0100 1011 1000.
- $(8)_{10}$ = $(008)_{10}$ = $(0000\ 0000\ 1000)_{BCD}$. The excess-3 equivalent of $(0000\ 0000\ 1000)_{BCD}$ = 0011 0011 1011.
- Subtracting-

```
  0100 1011 1000
- 0011 0011 1011
------------------------
  0001 0111 1101
```

Ex.5: Realize a four-input OR gate using two-input OR gates only.

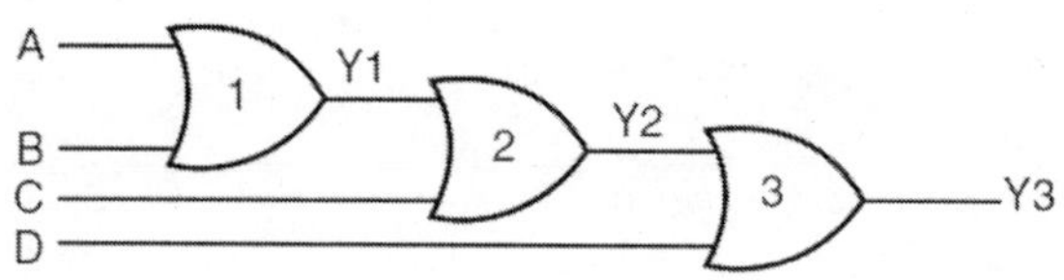

(a)

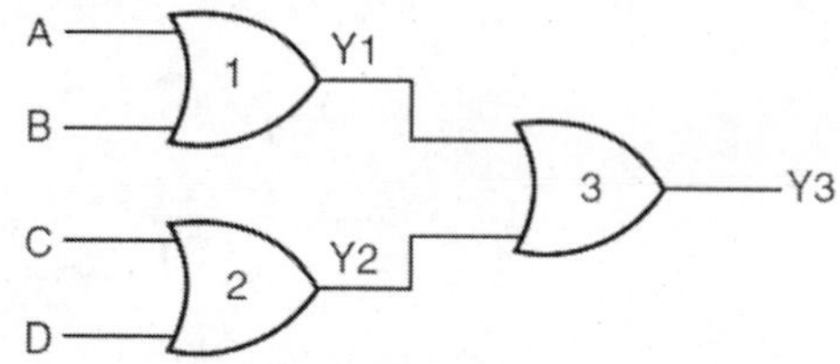

(b)

Solution: Fig. (a) shows an arrangement of two-input OR gates to work as a four-input OR gate. A, B, C, and D are logic inputs and Y3 is the output.

Fig. (b) shows another possible arrangement.

In the case of Fig. (a), the output of OR gate 1 is Y1 = (A+B). The second OR gate produces the output Y2 = (Y1+C) =(A+B+C). Similarly, the output of OR gate 3 is Y3 =(Y2+D) =(A+B+C+D).

In the case of Fig.(b), the output of OR gate 1 is Y1 =(A+B). The second OR gate produces the output Y2 =(C +D). Output Y3 of the third OR gate is given by(Y1+Y2) = (A+B+C +D).

Ex.6: Realize a four-input AND gate using two-input AND gates only.

Solution: Fig. below shows the implementation of a four-input AND gate using two-input AND gates. The output of AND gate 1 is Y1 = A.B. The second AND gate produces an output Y2 given by Y2 = Y1.C = A.B.C. Similarly, the output of AND gate 3 is Y = Y2.D = A.B.C.D.

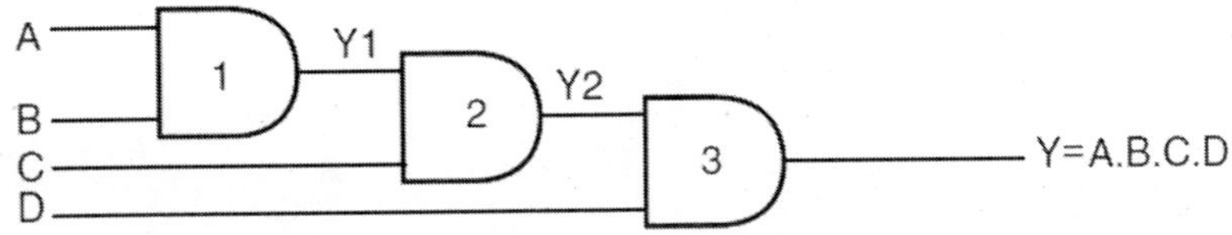

Ex. 7: Realize the followings-

(a) a four-input NAND gate using two-input AND gates and NOT gates;

(b) a three-input NAND gate using two-input NAND gates;

(c) a NOT circuit using a two-input NAND gate;

(d) a NOT circuit using a two-input NOR gate;

(e) a NOT circuit using a two-input EX-NOR gate.

Solution:

(a) The figure below shows the arrangement. The logic diagram is self-explanatory. The first step is to get a four-input AND gate using two-input AND gates. The output thus obtained is then complemented using a NOT circuit as shown.

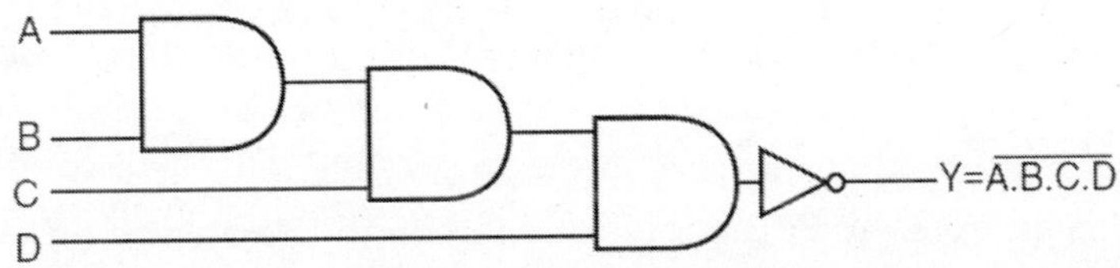

(b) The first step is to acquire a two-input AND from a two-input NAND. The output of the two-input AND gate and the third input then feed the inputs of another two-input NAND to get the desired output.

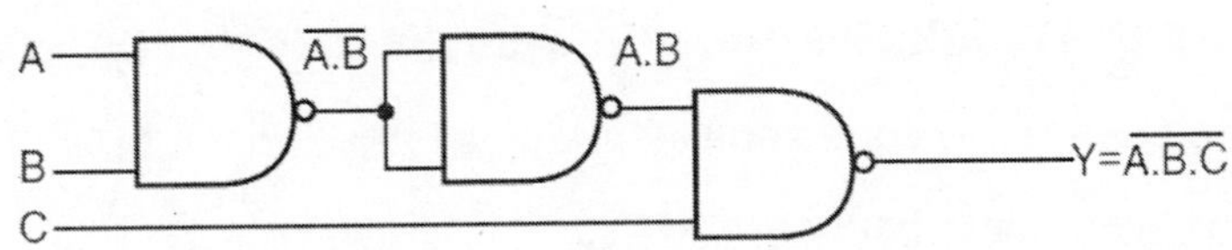

(c) Shorting the inputs of the NAND gives a one-input, one-output NOT circuit. This is because, a NAND gate, a similar set of inputs produces a logic LOW(0) output.

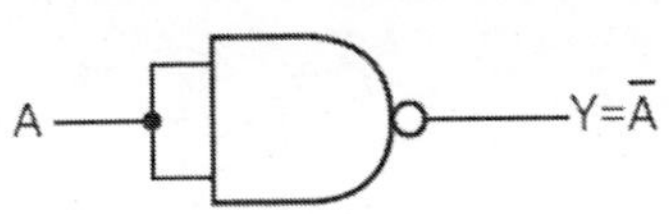

(d) Shorting the inputs of a NOR gate gives a NOT circuit. For a NOR gate, when all inputs are at logic zero, the output is HIGH(1) and vice-versa.

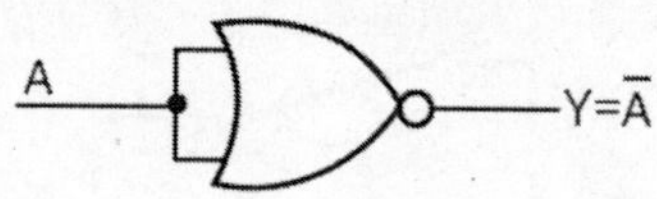

(e) In a two-input EX-NOR gate, if one of the inputs is permanently tied to a logic '0' level and the other input is treated as the input, then it behaves as a NOT circuit between input and output.

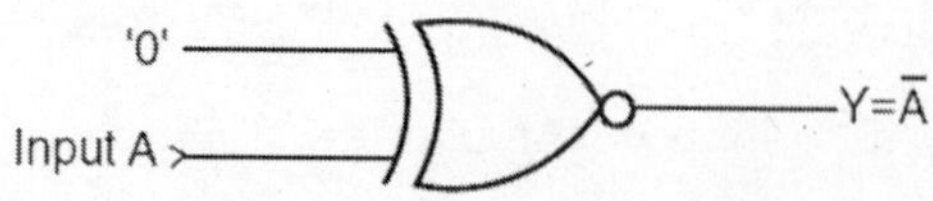

EXERCISE QUESTIONS

1. What do you understand by the 1's and 2's complements of a binary number?
2. Show that:

   (a) $(13A7)_{16} = (5031)_{10}$;

   (b) $(3F2)_{16} = (1111110010)_2$.
3. Distinguish between weighted and unweighted codes. Give two examples of both codes in brief.
4. What is an excess-3 BCD code? Illustrate with the help of an example, how the shortcoming of the 8421 BCD code is overcome in the excess-3 BCD code.
5. What is a Gray code? Why it is also called a binary-reflected code? Briefly discuss a few applications of a Gray code.
6. In 16-bit format, write the excess-3 equivalent codes for $(5)_{10}$, $(68)_{10}$, and $(467)_{10}$.
7. Determine the Gray code equivalent of $(10111)_2$ and the binary equivalent of the Gray code number 1010010.
8. Perform the binary division up to two binary places:

   (a) $(100.0001)_2 \div (10.1)_2$.

   (b) $(111001)_2 \div (1001)_2$.

   (c) $(111.001)_2 \times (1.11)_2$.
9. Perform the following operations using 2's complement arithmetic:

   (a) $(+43)_{10} - (-53)_{10}$.

(b) $(1ABC)_{16} + (1DEF)_{16}$.

(c) $(3E91)_{16} - (1F93)_{16}$.

10. What is the concept of a Universal gate? With suitable examples, explain why NAND and NOR gates are called universal gates.
11. Realize Ex-OR gate using NAND gate and NOR gates.
12. Draw the truth table of the logic circuit shown in the figure below-

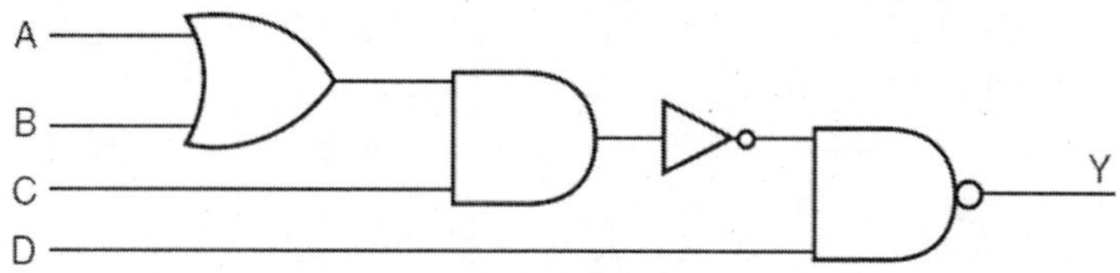

Self-Study Questions

1. How do you distinguish between positive and negative logic systems? Prove that an OR gate in a positive logic system is an AND gate in a negative logic system.
2. Give brief statements that would help one remember the truth table of AND, NAND, OR, NOR, EX-OR, and EX-NOR logic gate functions, irrespective of the number of inputs used.
3. What is meant by the radix or base of a number system? Briefly describe why hex representation is used for the addresses and the contents of the memory locations in the main memory of a computer.
4. What will be the range of decimal numbers that can be represented using a 16-bit 2's complement format?
5. In a number system, what decides (a) the place value or weight of a given digit and (b) the maximum numbers representable with a given number of digits?

Multiple Choice Questions

1. What is Digital Electronics?
   a) Field of electronics involving the study of digital signal
   b) Engineering of devices that digital signal
   c) Engineering of devices that produce a digital signal
   d) All of the mentioned
2. Which of the following is correct for Digital Circuits?
   a) Less susceptible to noise or degradation in quality
   b) Use transistors to create logic gates to perform Boolean logic
   c) Easier to perform error detection and correction with digital signal
   d) All of the mentioned
3. Which of the following is a type of digital logic circuit?
   a) Combinational logic circuits
   b) Sequential logic circuits
   c) Both a & b
   d) None of the mentioned
4. Which of the following codes is a sequential code?
   a) 8421 code
   b) 2421 code
   c) 5421 code
   d) 2441 code
5. Which of the following options correctly represents the characteristic of the Excess – 3 code?
   a) It is a reflexive as well as a sequential code
   b) It is a reflexive code but not a sequential code
   c) It is a sequential code but not a reflexive code
   d) It is neither a reflexive code nor a sequential code
6. The result "X + XY = X" follows which of these laws?
   a) Consensus law
   b) Distributive law
   c) Duality law
   d) Absorption law
7. Which of the following points is not correct regarding an Ex–NOR gate in Digital Electronics?
   a) It is a one-bit comparator
   b) It is a buffer
   c) It is a one–bit inverter
   d) It is a universal gate
8. Which gate is called the anti–coincidence and coincidence gate respectively?
   a) XNOR and XOR
   b) AND and OR
   c) OR and AND
   d) XOR and XNOR

9. Which of the following options represents the correct reduction of XYZ + XYZ?
a) 0
b) $\overline{Y}Z$
c) X + X
b) 2YZ

10. What determines the output from the combinational logic circuit in Digital Electronics?
a) Input signals from the past condition
b) Input signals at the present moment
c) Input signals from both past and present
d) Input signals expected in future

11. Any signed negative binary number is recognized by its ________
a) MSB
b) LSB
c) Byte
d) Nibble

12. If the decimal number is a fraction then its binary equivalent is obtained by ________ the number continuously by 2.
a) Dividing
b) Multiplying
c) Adding
d) Subtracting

13. The representation of an octal number (532.2)8 in decimal is ________
a) $(346.25)_{10}$
b) $(532.864)_{10}$
c) $(340.67)_{10}$
d) $(531.668)_{10}$

14. An important drawback of a binary system is ________
a) It requires a very large string of 1's and 0's to represent a decimal number
b) It requires a sparingly small string of 1's and 0's to represent a decimal number
c) It requires a large string of 1's and a small string of 0's to represent a decimal number
d) It requires a small string of 1's and a large string of 0's to represent a decimal number

15. On multiplication of (10.10) and (01.01), we get ____________
a) 101.0010
b) 0010.101
c) 011.0010
d) 110.0011

16. In boolean algebra, the OR operation is performed by which properties?
a) Associative properties
b) Commutative properties
c) Distributive properties
d) All of the Mentioned

17 A(A + B) = ?
a) AB

b) 1
c) (1 + AB)
d) A

18 DeMorgan's theorem states that _________
a) (AB)' = A' + B'
b) (A + B)' = A' * B
c) A' + B' = A'B'
d) (AB)' = A' + B

19 The code where all successive numbers differ from their preceding number by a single bit is _________
a) Alphanumeric Code
b) BCD
c) Excess 3
d) Gray

20 . How many AND gates are required to realize Y = CD + EF + G?
a) 4
b) 5
c) 3
d) 2

21 The NOR gate output will be high if the two inputs are __________
a) 00
b) 01
c) 10
d) 11

22 How many two-input AND and OR gates are required to realize Y = CD+EF+G?
a) 2, 2
b) 2, 3
c) 3, 3
d) 3, 2

23 How many two-input AND gates and two-input OR gates are required to realize Y = BD + CE + AB?
a) 3, 2
b) 4, 2
c) 1, 1
d) 2, 3

24 How many truth table entries are necessary for a four-input circuit?
a) 4
b) 8
c) 12
d) 16

25 Which input values will cause an AND logic gate to produce a HIGH output?
a) At least one input is HIGH
b) At least one input is LOW
c) All inputs are HIGH
d) All inputs are LOW

26 The AND function can be used to __________ and the OR function can be used to __________

a) Enable, disable
b) Disable, enable
c) Synchronize, energize
d) Detect, invert

27 If we use an AND gate to inhibit a signal from passing one of the inputs must be ____________

a) LOW
b) HIGH
c) Inverted
d) Floating

28 Which of the following combinations of logic gates can decode binary 1101?

a) One 4-input AND gate
b) One 4-input AND gate, one inverter
c) One 4-input AND gate, one OR gate
d) One 4-input NAND gate, one inverter

29 CMOS technology used in?

a) Microprocessor
b) Inverter
c) Digital logic
d) Both options a and c

MCQ Answer key:

1	**d**	9	**b**	17	**d**	25	**c**
2	**d**	10	**b**	18	**a**	26	**a**
3	**c**	11	**a**	19	**d**	27	**a**
4	**a**	12	**b**	20	**d**	28	**b**
5	**a**	13	**a**	21	**a**	29	**d**
6	**d**	14	**a**	22	**a**		
7	**a**	15	**c**	23	**a**		
8	**d**	16	**d**	24	**d**		

@@@@@@@@@

REFERENCES FOR FURTHER LEARNING

1) Robert Boylestad and Louis Nashelsky, **Electronic Devices, and Circuit Theory**,11th Edition, Pearson Education. ISBN:978-0-13-262226-4

2) Thomas Floyd, **Digital Fundamentals**, 11th Edition, Pearson Education, 2017, ISBN: 978-1-292-07598-3

3) David Bell, **Electronic Instrumentation and Measurements**, 2nd Edition, Prentice-Hall of India Pvt. Ltd.2003. ISBN:81-203-2360-2

4) A.K. Sawhney, **A Course In Electrical and Electronic Measurements and Instrumentation**, 2015, Dhanpat Rai & Co. (P) Limited, India. ISBN: 978-8177001006

CO AND PO ATTAINMENT TABLE

Course outcomes (COs) for this course can be mapped with the programme outcomes (POs) after the completion of the course and a correlation can be made for the attainment of POs to analyze the gap. After proper analysis of the gap in the attainment of POs necessary measures can be taken to overcome the gaps.

Table for CO and PO attainment

Course Outcomes	**Attainment of Programme Outcomes** *(1- Weak Correlation; 2- Medium correlation; 3- Strong Correlation)*											
	PO-1	**PO-2**	**PO-3**	**PO-4**	**PO-5**	**PO-6**	**PO-7**	**PO-8**	**PO-9**	**PO-10**	**PO-11**	**PO-12**
CO-1												
CO-2												
CO-3												
CO-4												
CO-5												
CO-6												
CO-7												

The data filled in the above table can be used for gap analysis.

INDEX

A

Absorptive law,237

AC diode model,35

Accelerometer,191

Active Filter using Op-Amp,133

Annulment law,237

Application of Op-Amp,127

Application of Zener diode,38

Applications,39

Associative law,237

Avalanche Brekdown,33

B

BCD-to-Binary Conversion,224

Bias Compensation,62

Binary Coded Decimal (BCD),223

Binary Codes,223

Binary Division, 230

Binary Number System,206

Binary-Gray Code Conversion,226

Binary-Hex and Hex-Binary Conversions,217

Binary-Octal and Octal-Binary Conversions,213

Binary-to-BCD Conversion,224

Binary-to-Decimal Conversions,210

Bipolar Junction Transistor (BJT),51

Block Diagram of Positive Feedback Amplifier,100

Breakdown Mechanisms,32

Bridge Rectifier,25

Buffer Amplifier or Voltage Follower,126

Built-in voltage of the P-N junction,6

By changing the area of overlapping plates,186
By changing the permittivity of the dielectric medium,186

C

Capacitive Transducer,184
Capacitive Tuning,108
Capacitor Filter,29
Characteristics of a Transducers,172
Characteristics of Zener diodes,38
Classification of Oscillator,101
Classification of Transducer,173
Close loop Configurations of an Op-Amp,122
CMOS,75
Colpitts Oscillator,106
Common Base Configuration (CB),55
Common Collector Configuration(CC),57
Common Emitter Configuration(CE),56
Commutative law,237
Comparison between BJT and FET,70
Complement law,238
Concept of Virtual ground,120
Crystal Oscillator,107
Current to voltage converter,127
Current-Series Feedback,96

D

D' Arsonval Movement,158
DC diode model,35
DC Load Line,58
DC Power Supply,16
Decimal Number System,203
Decimal-to-Binary Conversions,210
Decimal-to-Hexadecimal Conversions,212

Decimal-to-Octal Conversions,211
Differential Amplifier,121
Diode as a Rectifier,17
Diode as a switch,36
Diode capacitance,12
Diode Equivalent Circuit,35
Distributive law,237
D-MOSFET (n-channel),70

E

Effect of negative feedback on the gain and bandwidth,97
Electrical Equivalence Osillator,107
Electrical Transducers,171
Electrodynamic Frequency Meter,169
E-MOSFET (n-channel),73
Energy Band Diagram for a P-N junction,5
Energy band diagrams for P-N junction with external bias,11
Excess-3 Code,225

F

Factor influencing the choice of transducers,172
Feedback connection types,92
FET Parameters,69
Filters,27
For NPN transistors,54
For PNP transistors,54
Forward bias condition,8
Forward Bias,9
Frequency Meters,168
Full Wave Centre tapped rectifier,22

G

Galvanometer,164

General characteristics of a negative feedback amplifier,97

General characteristics of an Op-Amp,115

Generalized Expression for S,61.

Gray Code,226

Gray Code-Binary Conversion,228

H

Half wave Rectifier,17

Hall Effect Transducer,182

Hartley Oscillator,106

Hexadecimal Number System,207

Hexadecimal-to-Decimal Conversions,210

Hex-Octal and Octal-Hex Conversions,222

High pass Active filter,135

High pass Active filter,136

I

Idempotent law,238

Identity law,238

Important Specification of Op-Amp,118

Induction filter,27

Inductive Tuning,108

Input Impedance with feedback,94

Input impedance with feedback,94

Input offset Voltage(V_{io}),118

Input stage,116

Intermediate stage,116

Inverter (Sign Changer),126

Inverting Amplifier with Feedback,125

Inverting Amplifier,121

J

Junction Field effect transistors (JFET),64

L

LC Oscillator,105

Level Shifter,116

Light Emitting Diodes(LED),37

Liner Potentiometers,168

Liner variable differential transformer (LVDT),179

Low pass filter,134

M

Maximum Power Dissipation ($P_{Z\ max}$),38

Mixed tuning,108

Modes (Regions) of operation of BJT,52

MOSFET (Metal oxide semiconductor field effect transistor),70

Moving Iron Frequency Meter,168

Multi range Ohmmeter,164

N

Non-Inverting Amplifier with Feedback,122

Non-Inverting Amplifier,121

Number Representation in Binary System,208

O

Octal Number System,207

Octal-to-Decimal Conversions,210

Ohmmeter,162

OP-Amp as an Integrator,131

OP-Amp as a Subtractor,130

OP-Amp as Differentiator,132

Open Loop configuration of an Op-Amp,120
Operation of a JFET (n-channel),65
Oscillators,98
Output Impedance with feedback,95
Output offset voltage(V_{oo}),118
Output stage,116

P

Permanent Magnet Moving Coil Instrument,159
Photo Diode,38
Piezoelectric Transducer,188
P-N Junction with external bias,7
Position of Fermi level in doped Semiconductors,5
Potentiometer,166

R

RC phase shift oscillator,103
Reduction in frequency distortion,97
Reduction in nonlinear distortion and noise,97
Reverse Bias,10
Review of a P-N junction,3
Rotary Potentiometer,167

S

Seismic Transducer,190
Semiconductor band and Fermi energy levels,5
Sensistor Compensation,63
Sensors and Transducers,171
Series Ohmmeter,162
Shunt Ohmmeter,163
Sign-Bit Magnitude,208
Silicon Controlled Rectifiers,40
Special Purpose Diodes,37

Specification of Op-Amp,118
SRC Characteristics and Ratings,42
Static and dynamic resistance of a P-N junction diode,11
Strain Gauges,174
Summing, Scaling, and Averaging Amplifiers,127
Switching time of a Diode,15

T

Temperature Effect on Diode,14
The AND Gate,233
The Boolean algebra,236
The Combinational Circuits,238
The concept of feedback,90
The Decimal Equivalent,209
The Digital Arithmetic,229
The Digital Circuits,238
The EXCLUSIVE-NOR Gate,235
The EXCLUSIVE-OR Gate,234
The Field Effect Transistor (FET),64
The gain with feedback,93
The Logic Gates,232
The NAND Gate,234
The NOR Gate,234
The NOT Gate,233
The number system,203
The OR Gate,232
The Sequential Circuits,240
The Universal Gates,235
The V-I Characteristics,3
Thermal Stability,61
Thermistor Compensation,62
Thermistors,176
Transistor as Amplifier,63

Transistor biasing,59
Transistor Configuration,55
Transistor operation in linear mode,54

V

Vibrating-reed Frequency Meter,170
Vibrometer,191
Voltage-Series Feedback,96
Voltage-Shunt feedback,95

W

Wien Bridge Oscillator,104
Working Principal of Galvanometer,165

Z

Zener Breakdown,33
Zener Current(I_z),38
Zener Diode as a Voltage Regulator,31
Zener Diode,37
Zener Resistance (R_Z),38
Zener Voltage(V_z),38
Zero bias voltage,7